Clean Numerical Simulation

This book describes the basic principles of the Clean Numerical Simulation (CNS) proposed by the author in 2009, and several of its applications. Unlike conventional algorithms, CNS can give convergent chaotic trajectory in a sufficiently long interval of time, and whose numerical noise is much lower than the true physical solution P so that one can gain P accurately. Thus, CNS provides for the first time an ability to check the statistics stability of chaos leading to a completely new concept of "ultra-chaos" that has both trajectory instability and statistics instability, and thus is of a higher disorder.

Notably, it is impossible to repeat the experimental results of ultra-chaos even in statistical senses. However, the reproducibility of physical experiments forms a cornerstone of modern science. Thus, ultra-chaos reveals an incompleteness of the modern scientific paradigm. In addition, it reveals that statistics stability is a precondition for the use of conventional algorithms, including direct numerical simulation (DNS). In *Clean Numerical Simulation*, several conjectures and open problems are proposed, including a modified fourth Clay millennium problem.

Indeed, CNS opens the door to a "clean" numerical world of chaos and turbulence.

Advances in Applied Mathematics
Series Editor: Daniel Zwillinger

Introduction to Quantum Control and Dynamics
Domenico D'Alessandro

Handbook of Radar Signal Analysis
Bassem R. Mahafza, Scott C. Winton, Atef Z. Elsherbeni

Separation of Variables and Exact Solutions to Nonlinear PDEs
Andrei D. Polyanin, Alexei I. Zhurov

Boundary Value Problems on Time Scales, Volume I
Svetlin Georgiev, Khaled Zennir

Boundary Value Problems on Time Scales, Volume II
Svetlin Georgiev, Khaled Zennir

Observability and Mathematics
Fluid Mechanics, Solutions of Navier-Stokes Equations, and Modeling
Boris Khots

Handbook of Differential Equations, Fourth Edition
Daniel Zwillinger, Vladimir Dobrushkin

Experimental Statistics and Data Analysis for Mechanical and Aerospace Engineers
James Middleton

Advanced Engineering Mathematics with MATLAB®, Fifth Edition
Dean G. Duffy

Handbook of Fractional Calculus for Engineering and Science
Harendra Singh, H. M. Srivastava, Juan J. Nieto

Advanced Engineering Mathematics
A Second Course with MATLAB®
Dean G. Duffy

Quantum Computation
Helmut Bez and Tony Croft

Computational Mathematics
An Introduction to Numerical Analysis and Scientific Computing with Python
Dimitrios Mitsotakis

Delay Ordinary and Partial Differential Equations
Andrei D. Polyanin, Vsevolod G. Sorkin, Alexi I. Zhurov

Clean Numerical Simulation
Shijun Liao

https://www.routledge.com/Advances-in-Applied-Mathematics/book-series/CRCADVAPPMTH?
pd=published,forthcoming&pg=1&pp=12&so=pub&view=list

Clean Numerical Simulation

Shijun Liao

CRC Press
Taylor & Francis Group
Boca Raton London New York

CRC Press is an imprint of the
Taylor & Francis Group, an **informa** business
A CHAPMAN & HALL BOOK

First edition published 2024
by CRC Press
6000 Broken Sound Parkway NW, Suite 300, Boca Raton, FL 33487-2742

and by CRC Press
4 Park Square, Milton Park, Abingdon, Oxon, OX14 4RN

CRC Press is an imprint of Taylor & Francis Group, LLC

© 2024 J. Shijun Liao

ISBN: 978-1-032-28809-3 (hbk)
ISBN: 978-1-032-29023-2 (pbk)
ISBN: 978-1-003-29962-2 (ebk)

DOI: 10.1201/9781003299622

Typeset in CMR10
by KnowledgeWorks Global Ltd.

Publisher's note: This book has been prepared from camera-ready copy provided by the authors.

To my wife Shi Liu,
my daughter Fangzhou Liao,
and my granddaughter Ning Wu-Liao

Contents

Preface

Chaos dynamics [1–8] is widely regarded as one of the three greatest scientific revolutions in physics to have occurred in the twentieth century, comparable to quantum mechanics and Einstein's theory of relativity. The remarkable characteristic of chaos is its sensitivity dependence on initial conditions (SDIC), which was first discovered in 1890 by Poincaré [1], rediscovered in 1963 by Lorenz [5], and popularised through the name, "butterfly effect," whereby a hurricane in North America might be created by the flapping of wings of a distant butterfly in South America several weeks earlier. SDIC indicates the **trajectory instability** of chaos.

Lorenz [9, 10] was the first to discover that a chaotic trajectory exhibits sensitivity dependence not only on initial conditions but also on artificial factors (SDAFs) such as the numerical algorithm in single (or double) precision floating-point arithmetic, time-step, and so on. This, however, led to a heated debate [11–14] on the convergence and reliability of numerical simulations of chaotic systems because, unlike the initial condition, which has a purely physical meaning, the numerical algorithm, time-step, etc., are intrinsically *artificial*. But sensitivity dependence on artificial factors (SDAF) has no physical meaning whatsoever. One commentator even came to the rather pessimistic conclusion that "all chaotic responses are simply numerical noise and have nothing to do with the solutions of differential equations" [12]. The following question arises: *Can we produce a convergent trajectory of a chaotic system over a prescribed sufficiently long time interval?* In this book, I will give a positive answer to this fundamental question of chaos dynamics by introducing the basic principles of Clean Numerical Simulation (CNS), which I proposed in 2009, and illustrate several of its applications.

In practice, numerical noise caused by truncation error and round-off error is unavoidable and exponentially enlarges in chaotic systems due to the butterfly effect. Thus, for a conventional algorithm based on single (or double) precision floating-point arithmetic, a computer-generated chaotic simulation S' quickly becomes a mixture of the true physical solution $\mathcal{P}$ and false numerical noise δ', which are mostly of the *same* order of magnitude, i.e., $\delta' \sim \mathcal{P} \sim S'$, so that the chaotic simulation S' becomes badly polluted by numerical noise δ'. Despite this, it is widely accepted by the scientific community that the statistics of a chaotic system based on such a kind of mixture, i.e., $S' = \mathcal{P} + \delta'$, should be the *same* as those based on the true physical solution $\mathcal{P}$, i.e.,

$$\langle \mathcal{P} + \delta' \rangle = \langle \mathcal{P} \rangle, \quad \text{when } \delta' \sim \mathcal{P} \text{ mostly,}$$

where $\langle\rangle$ denotes a statistical operator. In other words, it is widely believed that, for a chaotic system, there should exist statistics stability despite trajectory instability. *Are the statistics of a chaotic system always stable?* This is another fundamental question of chaos dynamics. Unfortunately, there exist **no** rigorous proofs: $\langle\mathcal{P}+\delta'\rangle = \langle\mathcal{P}\rangle$ (when $\delta' \sim \mathcal{P}$ mostly) is only a **hypothesis** to the best of my knowledge. In this book, I will give a *negative* answer to this fundamental question by introducing a completely new concept of *ultra-chaos*, whose statistics are unstable, i.e., sensitive to small disturbances, and illustrating its widespread existence in chaotic systems. In addition, a completely new classification of chaos, into normal-chaos (with statistics stability) and ultra-chaos (with **statistics instability**), is described in Chapter 5 of this book. Obviously, ultra-chaos is of higher disorder than normal-chaos, since ultra-chaos has **both** trajectory instability and statistics instability!

Different from conventional algorithms in single (or double) precision, by means of CNS, one can gain a convergent chaotic simulation $\mathcal{S}$ over a sufficiently long interval of time $t \in [0, T_c]$, which can be used as a benchmark solution because its false numerical noise δ' is of much lower amplitude than that of its true physical solution $\mathcal{P}$, say, $|\delta'| \ll |\mathcal{P}|$, so that $\mathcal{S} = \mathcal{P} + \delta'$ is a very accurate approximation of $\mathcal{P}$, i.e., $\mathcal{S} \approx \mathcal{P}$. Here, T_c is called "the critical predictable time," which is one of the most important concepts in CNS. So, using this "clean" benchmark simulation $\mathcal{S}$, one can gain $\langle\mathcal{P}\rangle$ accurately, since $\langle\mathcal{P}\rangle \approx \langle\mathcal{S}\rangle$. Therefore, **for the first time**, CNS provides us with an ability to check the statistics stability of chaotic systems, i.e., $\langle\mathcal{P}+\delta'\rangle = \langle\mathcal{P}\rangle$, where the numerical simulation $\mathcal{S}' = \mathcal{P} + \delta'$ as a mixture is given by conventional algorithms using single (or double) precision, say, its false numerical noise δ' is mostly of the same order of magnitude as the true physical solution $\mathcal{P}$, and as mentioned above, $\langle\mathcal{P}\rangle$ can be accurately obtained by CNS, separately.

For an ultra-chaotic system, it is practically *impossible* to obtain any repeatable experimental/numerical results even in the statistical sense because the statistics of ultra-chaos is unstable, i.e., sensitive to small disturbances. Unfortunately, artificial and/or environmental disturbances are unavoidable and out of control. Thus, for an ultra-chaotic system, its reproducibility and/or replicability of physical experiments and/or numerical simulations are *inherently* impossible. However, the reproducibility and/or replicability of experiments form a **cornerstone** of modern science, ensuring that scientific laws remain invariant across time and space. Thus, the existence of ultra-chaos reveals a kind of **incompleteness** of the modern scientific paradigm, which, similar to the famous Gödel's incompleteness theorem in mathematical logic, indicates an inherent limitation in our current scientific research which might one day become a great challenge for humankind. Could such incompleteness shake the cornerstone of modern science? Could the existence of ultra-chaos lead to a crisis of confidence in scientific research? How should we understand and interpret numerical/experimental results of ultra-chaotic systems? How should we define "truth" in terms of ultra-chaos, and what kind of "truth" could an ultra-chaos tell us?

Note that the non-reproducibility of physical experiments is mentioned in the first volume (called "The Three Body Problem") of Liu Cixin's acclaimed science fiction trilogy, "Remembrance of Earth's Past" [15], which states: "they repeated the ultra-high-energy collision experiments again and again using the same conditions, but every time the result was different," and so "the development of physics stops." Thus, the hypothesis $\langle \mathcal{P} + \delta' \rangle = \langle \mathcal{P} \rangle$ (when $\delta' \sim \mathcal{P}$ mostly) is indeed both fundamental and highly important to humanity.

So, in Chapter 5 of this book, I propose the following two conjectures:

Conjecture 5.1 *Statistics instability*: There should exist certain dynamical systems, whose statistics are unstable, i.e., sensitive to small disturbances. In other words, $\langle \mathcal{P} + \delta' \rangle$ is not always equal to $\langle \mathcal{P} \rangle$, where $\langle \rangle$ denotes a statistical operator, $\mathcal{S}' = \mathcal{P} + \delta'$ is a numerical simulation of chaos obtained by conventional algorithms using single or double floating-point arithmetic, $\mathcal{P}$ denotes the true physical solution that can be accurately obtained by CNS, and δ' denotes false numerical noise, which all are mostly of the same order of magnitude, i.e., $\delta' \sim \mathcal{P} \sim \mathcal{S}'$.

Conjecture 5.2 *Incompleteness of modern scientific paradigm*: There should exist some dynamic systems with statistics instability, for which the current paradigm of modern science is invalid due to the loss of reproducibility and/or replicability of physical/numerical experiments.

It is hoped that rigorous proofs (or disproofs) of these two conjectures will be given in the near future. If true, a **paradigm shift** would be necessary in the further development of science and technology for humankind.

Note that "the critical predictable time" T_c is one of the most important concepts in the frame of CNS, since a CNS result $\mathcal{S}$ is convergent only within a *limited* time interval $t \in [0, T_c]$ and it is often computationally expensive to enlarge T_c for a chaotic system. Conversely, direct numerical simulation (DNS) [16, 17] involves *no* such concept at all because one can obtain a simulation $\mathcal{S}'$ by DNS over a time interval as long as would be desired: this, however, is based on the *hypothesis* that the corresponding statistics are stable, i.e., not sensitive to small disturbances including numerical noise, such that

$$\langle \mathcal{P} + \delta' \rangle = \langle \mathcal{P} \rangle, \quad \text{when } \delta' \sim \mathcal{P} \text{ mostly,}$$

where $\mathcal{P}$ is the true physical solution and δ' is the numerical noise, which is mostly of the same order of magnitude for conventional DNS algorithms in single (or double) precision floating-point arithmetic, say, $\delta' \sim \mathcal{P} \sim \mathcal{S}'$. Unfortunately, this hypothesis does *not* hold for ultra-chaotic systems, which widely exist in nonlinear dynamics, as illustrated in Chapter 5 of this book. Thus, the use of badly polluted DNS results has a *precondition*: statistics stability, corresponding to normal-chaotic systems. So, ultra-chaos, as a completely new concept, also reveals the precondition and limitation of DNS [16, 17].

Thus, I also propose several open questions in this book, including *a modified fourth Clay millennium problem* (see Chapter 6):

> **Open question 6.1** *The existence, smoothness, and statistics stability of the Navier–Stokes equation*: Can we prove the existence and smoothness of the solution of the Navier–Stokes equation with physically proper boundary and initial conditions, whose **statistics** are stable (or unstable), i.e., insensitive (or sensitive) to small disturbances?

The statistics stability of turbulence is very important in practice, because it is a precondition for the use of badly polluted simulations given by conventional algorithms in single (or double) precision including DNS [16, 17].

There are no rigorous proofs available to date about the statistics stability of the general N-body problem. So, I recommend the following open question:

> **Open question 7.1** *Statistics stability of N-body problem*: Can we rigorously prove (or disprove) the statistics stability $\langle \mathcal{P} + \delta' \rangle = \langle \mathcal{P} \rangle$ of the N-body problem with physically proper initial conditions when $\delta' \sim \mathcal{P}$ mostly, and N is a quite large positive integer such as $N = 1024^3$?

Such kind of statistics stability is very important for reliability of numerical simulations of the universe evolution. For details, see Chapter 7 of this book.

All of these aforementioned conjectures and open questions have very important meanings and should have far-reaching repercussions. Hopefully, CNS as a new tool and especially ultra-chaos as a completely new concept will open a door through which we can gain a deep understanding of chaotic systems, rigorously check the foundation of chaos dynamics, greatly enrich our knowledge of complicated systems including turbulent flows, and reveal inherent limitations on our paradigm of modern science.

I would like to express my sincere thanks to my former and current graduate students, Dr. Zhiliang Lin, Dr. Xiaoming Li, Mr. Tianzhuang Xu, Dr. Shijie Qin, Dr. Tianli Hu, Mrs. Yu Yang, Mr. Bo Zhang for their cooperation and hard work, and my collaborators Dr. Pengfei Wang at the Chinese Academy of Sciences and Prof.-Dr. Yipeng Ji and Prof.-Dr. Lipo Wang at the Shanghai Jiaotong University for their helpful discussions and cooperation. Particularly, I would like to express my sincere acknowledgements to my good friend, Prof.-Dr. Alistair Borthwick of the Universities of Edinburgh and Plymouth, for his friendship and encouragement over the past 20 years, and especially for his reading of the whole book and providing suggestions aimed at enhancing its readability and elegance.

Finally, I would like to express my pure-hearted thanks to my wife, Shi LIU, and my family for their love, patience, understanding, and encouragement.

The past three years have been hard for people all over the world. Hopefully, the COVID-19 pandemic and the ongoing wars across the world will come to an end as soon as possible so that all human beings on this beautiful planet can then enjoy health, security, peace, and love.

Shijun Liao

Shanghai, Spring 2023

About the Author

Shijun Liao is the Chun-Shen Distinguished Professor, Director of the State Key Laboratory of Ocean Engineering, and Dean of the School of Naval Architecture, Ocean and Civil Engineering at Shanghai Jiao Tong University, Shanghai, China. He holds a PhD from Shanghai Jiao Tong University and is a Cheung-Kong Distinguished Professor (Ministry of Education of China). His research is well known and includes topics such as nonlinear mechanics, gravity waves, turbulence, nonlinear dynamics, chaos, applied mathematics, analytic approximation method for highly nonlinear equations, reliable numerical simulations of chaotic systems and turbulence, and computer algebra methods in nonlinear mechanics. Awards and honors include: Shanghai Scientific Elite (2017), National Natural Science Award (2016), Shanghai Natural Science Award (2009), Shanghai Peony Natural Science Award (2009), Shanghai Excellent Teaching Award (2004), Thomson Reuters Highly Cited Researcher in Mathematics (2014, 2015, and 2016), Thomson Reuters Highly Cited Researcher in Engineering (2014), and World's Top 2% Scientists 2020 (Stanford University). He is the founder of the homotopy analysis method and the author of *Beyond Perturbation: Introduction to the Homotopy Analysis Method*, also published by CRC Press.

1

Introduction

This chapter describes the motivation behind Clean Numerical Simulation (CNS), briefly outlines its basic concepts, and presents applications that demonstrate the originality, validity, and great potential of CNS.

1.1 Motivation

The three-body problem can be traced back to Isaac Newton [18] in 1687 (see Proposition 66 of Book 1 and its 22 Corollaries and Propositions 25–35 of Book 3, *Mathematical Principles of Natural Philosophy*). The problem involves determining the motion of three masses, such as the Sun, Moon and Earth, using Newton's laws of motion and Newton's law of gravitational attraction. In Newton's own words (p153, Principia), " $\cdots$ to define these motions by exact laws admitting of easy calculation exceeds, if I am not mistaken, the force of any human mind." The three-body problem has influenced modern Science Fiction. In the first book of Liu Cixin's trilogy "Remembrance of Earth's Past" [15], a researcher called Wang Miao enters a trisolarian system where his world is influenced by the proximity of three suns. Wang says " $\cdots$ the reason why the sun's motion seems patternless is because our world has three suns. Under the influence of their mutually perturbing gravitational attraction, their movements are unpredictable – the three body problem. $\cdots$ This is a football game at the scale of the universe. The players are the suns, and our planet is the football."

In 1890, Poincaré [1] pointed out that the trajectories of three-body systems [19–23] are not integrable in general, and discovered the so-called "sensitivity dependence on initial condition" (SDIC), whereby a tiny variation in initial conditions might lead to considerable divergence in trajectory over a long period of time. In 1963, using numerical algorithms, Lorenz [2] rediscovered the SDIC of deterministic nonlinear dynamic systems by means of a computer, and helped give the SDIC a more popular name "the butterfly effect" whereby a hurricane in North America might be created by a flapping of the wings of a distant butterfly in South America several weeks earlier. The pioneering works by Poincaré [1] and Lorenz [2] are milestones in science, which introduced a new research field of nonlinear dynamics, later called chaos

DOI: 10.1201/9781003299622-1

(the name originally coined by Li and Yorke [3] in 1975). Nowadays chaos dynamics [4, 5, 7, 8, 24] has been widely regarded as one of the three greatest scientific revolutions in physics in the 20th century, comparable to quantum mechanics and Einstein's theory of relativity.

It is well-known [5, 7, 8, 24, 25] that due to the SDIC and the so-called butterfly effect [2], a tiny difference in initial condition of a chaotic system causes its trajectory to deviate exponentially. This characteristic of chaos is measured by the so-called maximum Lyapunov exponent. In theory, the maximum Lyapunov exponent of a chaotic dynamic system must be positive [5, 7, 8, 24, 25]. Today, high-performance computers are routinely used to produce numerical simulations of nonlinear differential/integral equations. Unfortunately, numerical noise arising from truncation and round-off errors is unavoidable in *all* numerical simulations. It is a common misconception that convergent (reliable) numerical simulations should invariably be obtained so long as the temporal and spatial discretizations are sufficiently fine. This is true for many dynamic systems with stable trajectories but unfortunately *not* for chaotic systems.

In 2006, Lorenz [10] numerically solved the following set of nonlinear ordinary differential equations

$$\begin{cases} \dot{X} = -Y^2 - Z^2 - aX + aF, \\ \dot{Y} = XY - bXZ - Y + G, \\ \dot{Z} = bXY + XZ - Z, \end{cases} \tag{1.1}$$

by means of first-order differencing with different values of time increment τ using double precision floating-point arithmetic*. Here X, Y, and Z are unknown functions of time, a, b, F, and G are physical parameters, and the dot indicates differentiation with respect to time t. Eqs. (1.1) were originally introduced to illuminate certain properties of atmospheric circulation. Lorenz [10] found that such a chaotic system is sensitive not only to initial conditions but also to the choice of *numerical algorithm*. As demonstrated in Figure 1.1, the maximum Lyapunov exponent, λ, of the chaotic model (1.1) for $a = 1/4$, $b = 4$, $F = 8$, and $G = 1$ invariably alternates between negative and positive values even when the time-step becomes rather small [10] for Eqs. (1.1) that are solved by first-order differencing in *double-precision* floating-point arithmetic. By progressively decreasing the time step τ, chaos is first observed in the range $0.0402 \leq \tau \leq 0.0435$ with positive λ, disappears with negative λ, re-emerges in the range $0.0344 \leq \tau \leq 0.0374$ with positive λ, disappears once more for smaller τ with negative λ, is observed once again when $\tau = 0.0028$ with positive λ, and then disappears for the smaller τ until $\tau = 0.00039$ with negative λ. This means that the basic characteristic of

---

*Please see https://en.wikipedia.org/wiki/Floating-point_arithmetic. In this book, single (double) precision floating-point arithmetic is often simplified by single (double) precision, if not mentioned.

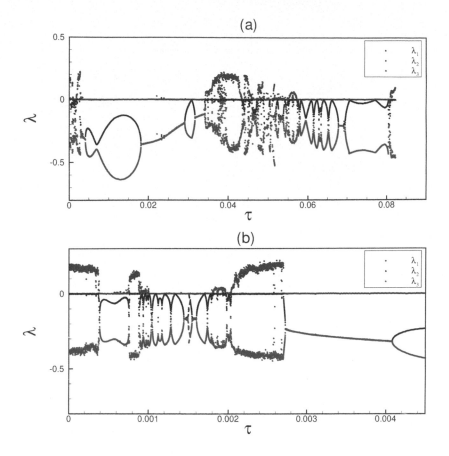

FIGURE 1.1

(a) Variation in maximum Lyapunov exponent λ with timestep τ obtained from numerical simulation of Eqs. (1.1) for $a = 1/4$, $b = 4$, $F = 8$ and $G = 1$ by means of first-order differencing for different values of time increment τ using double precision floating-point arithmetic. (b) Portion of (a) where $\tau < 0.0045$, horizontally stretched 20 times.

the numerical simulations alternates between non-chaotic and chaotic states, which are fundamentally quite different from each other. Using such numerical simulations, it turns out to be impossible to determine whether or not Eqs. (1.1) for $a = 1/4$, $b = 4$, $F = 8$, and $G = 1$ describe a chaotic system! Obviously, all the foregoing numerical simulations are unreliable because a convergent chaotic solution must be independent of any *artificial* factors such as time step, choice of numerical algorithm, etc.

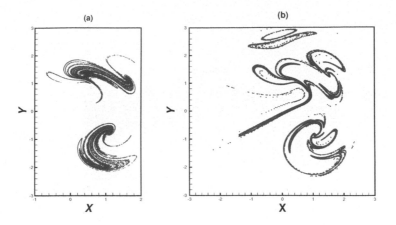

FIGURE 1.2

Intersection with plane $Z = 0$ of the attractor of Eqs. (1.1) for $a = 1/4$, $b = 4$, $F = 8$, and $G = 1$ predicted by first-order differencing in double-precision floating-point arithmetic with time increments: (a) $\tau = 0.037$; and (b) $\tau = 0.042$. Horizontal and vertical coordinates are X and Y, respectively.

Figure 1.2 shows cross-sections through the attractors of Eqs. (1.1) for $a = 1/4$, $b = 4$, $F = 8$, and $G = 1$, where the simulation is obtained using first-order differencing in *double-precision* floating-point arithmetic. Two cross-sections are plotted according to the time increments $\tau = 0.037$ and $\tau = 0.042$ that correspond to the prominent chaotic ranges in Figure 1.1(a). Although both attractors are strange, they are completely *different*. Rather unexpectedly, Figure 1.2(a) maps very neatly onto the interior empty spaces in Figure 1.2(b). The majority of states encountered when $\tau = 0.037$ appear to be avoided when $\tau = 0.042$. Obviously, from a physical viewpoint, the true solution of Eqs. (1.1) should be *independent* of the time increment τ (which is an *artificial* factor used in the numerical algorithm). Thus, both numerical simulations given by $\tau = 0.037$ and $\tau = 0.042$ are unreliable. The example nicely illustrates "the sensitivity dependence on numerical algorithms" (SDNA) of chaotic dynamic systems. SDNA was reported more than 40 years after Lorenz rediscovered SDIC [2] and is a basic characteristic of chaotic systems that is as fundamental as SDIC. To date, the SDNA of chaos has unfortunately been overlooked by most researchers.

Over the years, similar numerical phenomena have been rediscovered and/or confirmed. For example, Teixeira *et al.* [11] investigated the time-step sensitivity of three nonlinear atmospheric models by means of conventional algorithms in *double-precision* floating-point arithmetic, but came to the rather pessimistic conclusions that "for chaotic systems, numerical convergence cannot be guaranteed *forever*", and that "for regimes that are not fully chaotic, different time-steps may lead to different model climates and even different regimes of the solution."

In the field of numerical methods, it is widely believed that symplectic integrators [26–31] can provide more accurate simulations than other approaches because symplectic integrators can guarantee conservation of important physical quantities such as momentum, energy, etc. However, in 2015, Hoover *et al.* [14] examined Lyapunov instability by comparing numerical simulations of a chaotic Hamiltonian system given by two Runge-Kutta schemes and five symplectic integrators in *double-precision* floating-point arithmetic. Hoover *et al.* [14] concluded that "*all* numerical methods are susceptible to Lyapunov instability, which severely limits the *maximum time* for which chaotic solutions can be accurate", although "all of these integrators conserve energy almost perfectly" and "they also reverse back to the initial conditions even when their trajectories are inaccurate." As reported by Hoover *et al.* [14], "the advantages of higher-order methods are lost rapidly for typical chaotic Hamiltonians", and "there is little distinction between the symplectic and the Runge-Kutta integrators for chaotic problems, because *both* types lose accuracy at the *same* rate, determined by the maximum Lyapunov exponent." According to Hoover *et al.* [14], modifications to numerical algorithms *alone cannot* lead to better simulations of chaos (noting that *double-precision* floating-point arithmetic was used by Hoover *et al.*).

To examine the negative conclusions of Hoover *et al.* [14], let us consider the motion of a star orbiting in a plane about the galactic centre, governed by the so-called Hénon-Heiles Hamiltonian system of equations

$$\begin{cases} \ddot{x}(t) = -x(t) - 2x(t)y(t), \\ \ddot{y}(t) = -y(t) - x^2(t) + y^2(t). \end{cases} \tag{1.2}$$

Here, the Hamiltonian is the total energy, i.e.,

$$H = T(\dot{x}, \dot{y}) + V(x, y),$$

where

$$T(\dot{x}, \dot{y}) = \frac{1}{2}\left(\dot{x}^2 + \dot{y}^2\right), \qquad V(x, y) = \frac{1}{2}\left(x^2 + y^2 + 2x^2y - \frac{2}{3}y^3\right)$$

are kinetic energy and potential energy, respectively. As pointed out by Sprott [8], the solution is chaotic for certain initial conditions, such as

$$x(0) = \frac{14}{25}, \quad y(0) = 0, \quad \dot{x}(0) = 0, \quad \dot{y}(0) = 0. \tag{1.3}$$

Chaotic orbits of the Hénon-Heiles system (1.2) for initial condition (1.3) in the interval [0, 1000] are calculated by means of the 4th-order symplectic integrator in *double-precision* floating-point arithmetic using three different

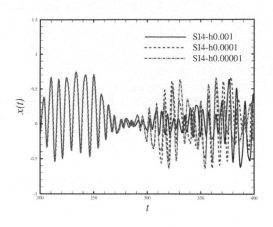

FIGURE 1.3

Simulations of displacement $x(t)$ in the Hénon-Heiles system (1.2) and (1.3) given by a fourth-order symplectic integrator using double-precision floating-point arithmetic for three values of time-step h. Solid line: $h = 0.001$; dashed line: $h = 0.0001$; dash-dotted line: $h = 0.00001$.

time-steps $\Delta t = h = 0.001, 0.0001$, and 0.00001. In all the simulations, the relative energy error given by the 4th-order symplectic integrator remains small over the whole interval $[0, 1000]$, i.e., less than 10^{-12}. However, as shown in Figure 1.3, the different orbits rapidly diverge from each other: the two orbits given by the symplectic integrator with $h = 0.001$ and $h = 0.0001$ separate from each other at about $t = 280$, and the two orbits obtained with $h = 0.0001$ and $h = 0.00001$ diverge from each other at about $t = 310$. Therefore, none of these trajectories given by the symplectic integrators are reliable in $t \in [0, 1000]$, even if the Hamiltonian (i.e., total energy of the system) is conserved quite well. These results confirm the negative conclusion of Hoover *et al.* [14].

Let us now consider another Hamiltonian system, the famous three-body problem governed by Newton's law of gravitational attraction and Newton's laws of motion. The body accelerations are expressed by

$$\ddot{x}_{k,i} = \sum_{j=1, j \neq i}^{3} m_j \frac{(x_{k,j} - x_{k,i})}{R_{i,j}^3}, \quad k = 1, 2, 3, \tag{1.4}$$

where $\mathbf{r}_i = (x_{1,i}, x_{2,i}, x_{3,i})$ and m_i denote the position and mass of the ith body $(i = 1, 2, 3)$, and

$$R_{i,j} = \left[\sum_{k=1}^{3} (x_{k,j} - x_{k,i})^2 \right]^{1/2}.$$

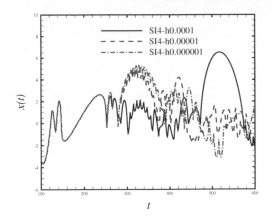

FIGURE 1.4

The x position of Body-1 in the three-body problem defined by Eqs. (1.4) and (1.5), obtained using a fourth-order symplectic integrator in double-precision floating-point arithmetic, for three values of time-step h. Solid line: $h = 10^{-4}$; dashed line: $h = 10^{-5}$; dash-dotted line: $h = 10^{-6}$.

Without loss of generality, we consider the case $m_1 = m_2 = m_3 = 1$ and the initial condition

$$\begin{cases} \mathbf{r}_1 = (1/10, 0, -1), \mathbf{r}_2 = (0, 0, 0), \mathbf{r}_3 = (0, 0, 1), \\ \dot{\mathbf{r}}_1 = (0, -1, 0), \dot{\mathbf{r}}_2 = (1, 1, 0), \dot{\mathbf{r}}_3 = (-1, 0, 0), \end{cases} \tag{1.5}$$

corresponding to a chaotic three-body system [8].

The fourth-order symplectic integrator in *double-precision* floating-point arithmetic is used to determine numerically the chaotic orbits of the three-body system (1.4) and (1.5) in the interval $t \in [0, 1000]$ for three values of time-step $\Delta t = h = 10^{-4}, 10^{-5}$, and 10^{-6}. Given that the three-body problem describes a Hamiltonian system, its total energy must be conserved for the simulation to be reliable. In this case, the deviation in total energy of the three-body system given by the symplectic integrator in *double-precision* floating-point arithmetic is indeed rather small within the *whole* interval $[0,1000]$, remaining at a level less than 10^{-8}. Unfortunately, even this *cannot* guarantee the convergence of the chaotic trajectory of the three-body system. As shown in Figure 1.4, the x position of Body-1 determined for a time-step $h = 10^{-4}$ diverges at $t \approx 260$ from that obtained for $h = 10^{-5}$, and the x positions obtained for $h = 10^{-5}$ and $h = 10^{-6}$ depart from each other at $t \approx 310$. Therefore, the long-term evolution of chaotic orbits in the three-body system (1.4) and (1.5) obtained using a fourth-order symplectic integrator in *double-precision* floating-point arithmetic are *unreliable* when $t > 310$. The foregoing supports Hoover *et al.*'s negative conclusion [14] that even symplectic integrators *cannot* guarantee the convergence and reliability of trajectories in chaotic dynamic systems over a long interval of time (when *double-precision* floating-point arithmetic is used in the computations).

It should be emphasized that it can also be rather difficult to achieve accurate numerical predictions for certain dynamic systems without violating Lyapunov stability. For example, let us consider the Lorenz equation [2]

$$\begin{cases} \dot{x} = -\sigma x + \sigma y, \\ \dot{y} = rx - y - x\,z, \\ \dot{z} = x\,y - bz, \end{cases} \tag{1.6}$$

where $x(t), y(t)$, and $z(t)$ are unknown variables; σ, b, and r are physical parameters; and the dot denotes differentiation with respect to time t. Let $\sigma = 10$ and $b = 8/3$. Theoretically speaking, when $1 < r < 24.74$, the long-term solution of the Lorenz equations should finally tend to one of two *stable* fixed points given by

$$C(\sqrt{b(r-1)}, \ \sqrt{b(r-1)}, \ r-1)$$

and

$$C'(-\sqrt{b(r-1)}, \ -\sqrt{b(r-1)}, \ r-1).$$

However, Li *et al.* [32] studied the sensitive dependence of the fixed point for $r = 22$ on the time-step h using many explicit/implicit numerical approaches (such as Euler's method, Runge-Kutta methods of orders from 2 to 6, Taylor series methods of orders from 2 to 10, Adams methods of orders from 2 to 6, etc.) in *double-precision* floating-point arithmetic. Li *et al.* found that long-term numerical simulations were rather sensitive to the value of the time-step $\Delta t = h$, with the results always fluctuating between the two fixed points, no matter how small the time-step h. Thus, Li *et al.* also came to the rather pessimistic conclusion that a "numerical solution obtained by *any* step size is *unrelated* to exact solution" [32].

To further investigate Li *et al.*'s [32] negative conclusion, we now solve the Lorenz equation (1.6) for $\sigma = 10, b = 8/3$, and $r = 23$, and the exact initial condition

$$x(0) = 5, \quad y(0) = 5, \quad z(0) = 10 \tag{1.7}$$

using the fourth-order Runge-Kutta method in *double-precision* floating-point arithmetic. For $r = 23 < 24.74$, the corresponding long-term numerical trajectory should in theory eventually tend to one of two stable fixed points, $x = \pm 4\sqrt{11/3}$. We find that the final steady results are indeed rather sensitive to the time-step $\Delta t = h$, as shown in Figure 1.5; no matter how small the time-step h, the results always fluctuate between the two fixed points $x = \pm 4\sqrt{11/3}$. This confirms Li *et al.*'s [32] pessimistic conclusion.

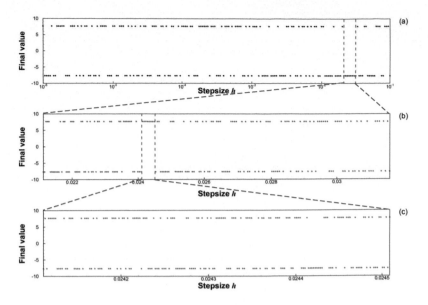

FIGURE 1.5
Final value of $x(t)$ versus step size h from simulations of the Lorenz equations
(1.6) for $\sigma = 10, b = 8/3$, and $r = 23$ subject to the initial condition (1.7),
obtained using a 4th-order Runge-Kutta method in double-precision floating-
point arithmetic. The step size h varies from (a) 10^{-6} to 10^{-1}, (b) 2.11349×10^{-2} to 3.16228×10^{-2}, and (c) 2.41107×10^{-2} to 2.45086×10^{-2}.

How to reduce the truncation error? For example, let us consider the
Mth-order Taylor expansion method [33–36], given by

$$f(t + h) = f(t) + \sum_{m=1}^{M} \frac{f^{(m)}(t)}{m!} h^m + R_T, \qquad (1.8)$$

where $f(t)$ is a real function, $f^{(m)}(t)$ denotes its mth-order derivative with
respect to t, and R_T denotes the truncation error[†]. Consider the Lorenz equation (1.6) for $\sigma = 10, b = 8/3, r = 23$, initial condition (1.7), and time step
$\Delta t = h = 0.01$. Obviously, if the time step h is smaller than the convergence radius, then the larger the order M of Taylor series, the smaller the
truncation error R_T. To increase computational efficiency, we use parallel
processing, with np denoting the number of processors. Surprisingly, when
double-precision floating-point arithmetic is used, different final steady results
are obtained using the high-order Taylor expansion method even when using

---

[†]https://en.wikipedia.org/wiki/Truncation_error

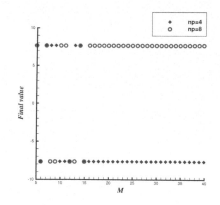

FIGURE 1.6

Final steady value of $x(t)$ from simulations of the Lorenz equations (1.6) versus order M for $\sigma = 10, b = 8/3$, and $r = 23$ subject to the initial condition (1.7), obtained using a truncated Mth-order Taylor's expansion method in double-precision floating-point arithmetic, computed on a Thinkpad L440 laptop computer with Intel Core i7-4712MQ for different numbers np of processors. Diamonds, $np = 4$; circles, $np = 8$.

the same laptop (Thinkpad L440 with Intel Core i7-4712MQ) but *different* numbers of processors, as shown in Figure 1.6. Even if the order (M) of the Taylor expansion method is rather high, e.g. $M = 200$, corresponding to a rather small truncation error, we nevertheless obtain different final steady results using *different* numbers of processors in the *same* laptop, as shown in Figure 1.7.

Oddly, when using the high-order Taylor series method in conjunction with *double-precision* floating-point arithmetic for the same initial condition, different final steady results (see Figure 1.8) are obtained using the *same* number of processors np but *different* types of computer (i.e., Thinkpad L440 laptop computer with Intel Core i7-4712MQ and Tianhe-II supercomputer, Guangzhou, China). This artificial uncertainty within the numerical simulations *cannot* be avoided even when using a rather high-order Taylor series method, as shown in Figure 1.9. The results indicate that reducing the truncation error *alone* cannot prevent artificial uncertainty in numerical simulations of many nonlinear systems.

Why? Is something wrong? There is an analogy with Shakespeare's famous soliloquy where Prince Hamlet says "To be, or not to be, that is the question!"

An initial condition has physical meaning. This is exemplified by sensitivity dependence on initial condition (SDIC), discovered by Poincaré and later called the butterfly effect by Lorenz at the Annual Meeting of the American Association for the Advancement of Science in 1972. However, choice

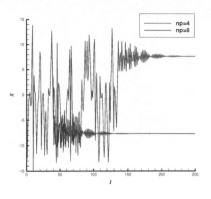

FIGURE 1.7
Predicted $x(t)$ time series from Lorenz equations (1.6) for $\sigma = 10, b = 8/3$, and $r = 23$ subject to the initial condition (1.7), obtained using a 200th-order Taylor's expansion method (i.e., $M = 200$) in double-precision floating-point arithmetic, computed on a Thinkpad L440 laptop computer with Intel Core i7-4712MQ for different numbers np of processors. Red line, $np = 4$; black line, $np = 8$.

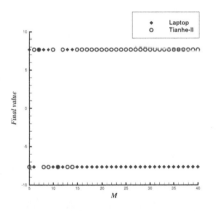

FIGURE 1.8
Predicted final steady value of $x(t)$ from Lorenz equations (1.6) for $\sigma = 10, b = 8/3$, and $r = 23$ subject to the initial condition (1.7) versus order M, obtained by means of a truncated Mth-order Taylor's expansion in double-precision floating-point arithmetic, computed using the same number of processors ($np=4$) but different computers. Diamonds, Thinkpad L440 laptop computer with Intel Core i7-4712MQ; circles, Tianhe-II supercomputer.

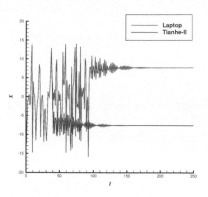

FIGURE 1.9

Predicted $x(t)$ time series from Lorenz equations (1.6) for $\sigma = 10, b = 8/3$, and $r = 23$ subject to the initial condition (1.7), obtained by means of a 200th-order Taylor's expansion method in double-precision floating-point arithmetic using the same number of processors ($np = 4$) but different computers. Red line, Thinkpad L440 laptop computer with Intel Core i7-4712MQ; black line, Tianhe-II supercomputer.

of numerical algorithm, number of processors, type of computer, etc., are all *artificial* factors, which lack physical meaning in the context of the nonlinear system being simulated. In fact, the solution of a mathematical model is completely determined by initial/boundary conditions and physical parameters and should have nothing to do with any artificial factors, such as choice of numerical algorithm, type of computer, etc. So, unlike SDIC, sensitivity dependence on numerical algorithm (SDNA) has *no* physical meaning, and therefore must be abandoned in theory.

The foregoing numerical phenomena cast a shadow that leads to serious doubt regarding the reliability of numerical simulations of chaos. Naturally, intense argument has been unavoidable [12,13]. For example, Yao and Hughes [12] commented that "all chaotic responses are simply numerical noise and have nothing to do with the solutions of differential equations." Yao [37] even suggested that "reports of computed non-periodic solutions of chaotic differential equations are simply consequences of unstably amplified truncation errors, and are not approximate solutions of the associated differential equations."

It should be emphasized that for all traditional algorithms there always exists artificial numerical noise caused by truncation and round-off errors[‡].

[‡]https://en.wikipedia.org/wiki/Round-off_error

For a chaotic system, such numerical noise grows exponentially due to the butterfly-effect so that a computer-generated simulation S' quickly becomes a mixture of the true physical solution $\mathcal{P}$ and the false numerical noise δ', which are mostly of the same order of magnitude, say, $\delta' \sim \mathcal{P} \sim S'$. Despite this well-known fact, numerical simulations involving such mixtures as $S' = \mathcal{P} + \delta'$ are widely used in practice to calculate the statistics of chaotic systems, because it is widely believed that, although trajectories of chaos are unstable, i.e., sensitive to small disturbance, its statistical results should be stable, say,

$$\langle \mathcal{P} + \delta' \rangle = \langle \mathcal{P} \rangle, \qquad \text{when } \delta' \sim \mathcal{P} \text{ mostly}, \tag{1.9}$$

where $\langle \rangle$ denotes a statistical operator, $\mathcal{P}$ denotes the true physical solution, and δ' denotes the false noise caused by artificial and/or environmental disturbances. Unfortunately, to the best of the author's knowledge, (1.9) has *not* been proved rigorously to date, and so is only a *hypothesis*!

What happens if hypothesis (1.9) is invalid? In such a case, the invalidity of (1.9) implies that the statistical results are *unstable*, i.e., sensitive to small disturbances. Obviously, it is practically impossible to repeat any experimental (or numerical) results of such kinds of unstable systems even in the statistical sense, because small artificial/environmental disturbances are unavoidable and out of control in general. For such kinds of unstable systems, neither experimental results nor numerical simulations can be repeated even in the statistical sense. However, reproducibility is a cornerstone of modern science! Therefore, the invalidity of hypothesis (1.9) reveals a kind of incompleteness of modern scientific paradigm and thus might shake its foundation, which, similar to Gödel's incompleteness theorem, indicates potential inherent limitations in our scientific researches and might eventually pose a major challenge for human beings in the future[§]. Thus, it is indeed a very serious problem whether hypothesis (1.9) is valid or not.

The foregoing has formed two dark clouds over the blue sky of chaos dynamics, which present scientists with a great challenge. From a theoretical perspective, we must answer the following two serious questions:

1. Is it possible to gain *convergent* (reliable) numerical simulations of chaos over a sufficiently long interval of time?

2. Do artificial and/or environmental disturbances influence numerical simulations of chaos in a *statistical* sense?

Thus, a new numerical strategy is required to obtain reliable, convergent numerical simulations of chaotic dynamic systems, called "clean" results,

---

[§]Non-reproducibility of physical experiments is mentioned in Liu Cixin's famous science fiction trilogy "Remembrance of Earth's Past" [15] in which it is stated that: "they repeated the ultra-high-energy collision experiments again and again using the same conditions, but every time the result was different", and as a result, many scientists believe that "the laws of physics are not invariant across time and space."

which are *insensitive* to *any* human/artificial factors such as choice of time step, numerical algorithm, number of processors, type of computer, and so on. Here, a "clean" numerical simulation is one where numerical deviations (denoted by δ') caused by artificial factors remain *negligible* over a prescribed, long interval of time $t \in [0, T_c]$, compared to the true physical solution $\mathcal{P}$, say, $|\delta'| \ll |\mathcal{P}|$ in mathematics, so that the chaotic numerical simulation $\mathcal{S}$ is very close to the true physical solution $\mathcal{P}$ in $t \in [0, T_c]$ and thus $\langle \mathcal{S} \rangle$ is a very good approximation of $\langle \mathcal{P} \rangle$. Obviously, such a "clean" chaotic simulation, which is convergent over a prescribed long interval of time $t \in [0, T_c]$, can be used as a benchmark solution to investigate the statistics stability of chaotic dynamic systems by checking the validity of hypothesis (1.9). In this way, we can clearly answer the two questions posed above. This provides the motivation behind the present monograph.

1.2   A Brief Description of Clean Numerical Simulation

Throughout the history of science, challenges and opportunities almost invariably coexist. An intense debate [12, 13, 37] stirred the author to examine the influence of artificial and/or environmental small disturbances on chaotic trajectories and related statistical results, and, hence, to put forward a new numerical strategy for modelling chaotic systems called "Clean Numerical Simulation" (CNS) [38–58], which could sweep away the two dark clouds hanging over chaos dynamics mentioned at the end of Section 1.1.

The examples described in Section 1.1 demonstrate that numerical simulations of chaotic dynamic systems are sensitive *not only* to initial conditions *but also* to certain artificial factors, such as time step, numerical algorithm, number of processors, type of computer, etc. Theoretically speaking, these *artificial* factors help generate truncation error and/or round-off error. Sensitivity dependence on initial condition (SDIC) of chaos has physical meaning. But, sensitivity dependence on artificial factors has *no* physical meaning whatsoever. Obviously, numerical simulations with sensitivity dependence on artificial factors are *not* reliable.

It should be emphasized that *double-precision* floating-point arithmetic was used in computing all the preceding examples of the numerical simulation of chaotic systems. By definition, this causes round-off error to enter the numerical simulations. Round-off error, also called rounding error, is the difference between the calculated approximation of a number and its exact mathematical value (which in theory can have an infinite number of digits). Round-off error is especially important when using *finite* digits to represent real numbers in numerical analysis. Of course, some level of round-off error is *unavoidable* in numerical simulations by means of a computer!

As pointed out by Monniaux [59], even when using the *same* programs with the *same* compiler to solve exactly the *same* expression with the *same* values for the *same* variables, etc., simulations on different working platforms may exhibit subtle differences with respect to floating-point computations. Thus, even the choice of number of processors operating on the same computer, or the choice of computer with the same number of processors, can generate rather tiny differences in round-off error, which unfortunately grow almost exponentially (due to the butterfly effect of chaos) such that completely different numerical simulations might be obtained after sufficient time has elapsed! In practice, the growth of round-off error can lead to very serious problems, such as the failure of a military system [60] on February 25, 1991, when a fixed-point round-off error in the radar tracking system of an MIM-104 Patriot missile battery prevented it from intercepting an incoming Scud missile in Dhahran, Saudi Arabia, contributing to the death of 28 soldiers and injuring a further 98. This highlights the profound influence of round-off error on reliable simulation of nonlinear dynamic systems.

Double-precision floating-point arithmetic was used in all the examples mentioned in Section 1.1, thus introducing round-off error into the numerical simulations. Unfortunately, for a chaotic system, the butterfly effect can cause even a tiny disturbance caused by the round-off error to increase exponentially to macroscopic level. This partly explains why many researchers (see e.g. [10–12, 14, 32]) were unable to obtain convergent, reliable simulations of chaos over a long time interval. Note that background numerical noise is determined by the maximum of the round-off error and the truncation error. Thus, when double-precision floating-point arithmetic is used, there is no point in trying to reduce the truncation error *alone*, as shown in Figures 1.7 and 1.9, where even a 200th-order Taylor expansion is utilised and the truncation error is much smaller than the round-off error! So, in order to achieve a reliable chaotic simulation over a sufficiently long time interval, we *must* abandon double-precision floating-point arithmetic. It is indeed a pity that the influence of round-off error on the reliability of numerical simulations of chaos did not attract enough attention in the past: as a consequence, double-precision floating-point arithmetic is routinely used in numerical simulations of chaos nowadays.

As shown in Figure 1.4, the deviations between numerical simulations obtained using three different time-steps $h = 10^{-4}$, $h = 10^{-5}$, and $h = 10^{-6}$ (with double-precision floating-point arithmetic) are negligible at early time in the interval $t \in [0, 280]$, then quickly increase, and remain very distinct after $t > 310$. Note that, after $t > 310$, the deviations in the three numerical simulations are of the same order of magnitude, and so are at the macroscopic level. In other words, for $t > 310$, these chaotic simulations S' given by conventional numerical algorithms using double-precision floating-point arithmetic are a *mixture* of the true physical solution P and false numerical noise δ', which are mostly of the *same* order of magnitude, say, $\delta' \sim P \sim S'$. Thus, for numerical simulations S of a chaotic system, a critical time T_c should exist below which the false numerical noise δ' is negligible with respect to the true physical solution $\mathcal{P}$, i.e., $|\delta'| \ll |\mathcal{P}|$, and thus can be ignored (i.e., $\mathcal{S} \approx \mathcal{P}$) for $t \in [0, T_c]$. For $t > T_c$, the false numerical noise δ' rapidly grows to the same order of magnitude as the true physical solution $\mathcal{P}$ (e.g., as evident in Figure 1.4), say, $\delta' \sim \mathcal{P}$. Here, T_c is called "the critical predictable time." For example, in Figure 1.4, $T_c \approx 280$ for $h = 10^{-4}$ and $T_c \approx 310$ for $h = 10^{-5}$. In other words, a numerical simulation is "clean" for $t \in [0, T_c]$, but becomes "dirty" for $t > T_c$, i.e., being badly contaminated by the false numerical noise δ'. In practice, numerical simulations of chaotic systems given by traditional algorithms in double-precision floating-point arithmetic often have rather small T_c. For example, chaotic numerical simulations of the Lorenz equation obtained using double-precision floating-point arithmetic are often convergent over a rather small interval $t \in [0, 32]$, which is too short for practical purposes. Therefore, it is important to greatly increase the critical predictable time T_c, which is a key tenet underpinning CNS, and then also to investigate the influence of false numerical noise on the statistics of chaos, which is important from both theoretical and practical perspectives.

CNS is based on the following well established phenomenon: in the numerical simulation of a chaotic dynamical system, the deviation (on average) from the true physical solution rises *exponentially* to become macroscopic at $t = T_c$, where T_c is called the critical predictable time. In other words, the average deviation of the simulation from the true physical solution $\mathcal{P}$ at time t is given by,

$$\mathcal{E}(t) = \mathcal{E}_0 \exp(\kappa t), \qquad t \in [0, T_c], \qquad (1.10)$$

where $\mathcal{E}_0$ denotes the level of background numerical noise (determined by the maximum of the truncation and round-off errors), and $\kappa > 0$ is the noise-growth exponent. In theory, the critical predictable time T_c may be determined by defining a critical level $\mathcal{E}_c$ of the simulation deviation from the true physical solution $\mathcal{P}$, such that

$$T_c = \frac{1}{\kappa} \ln \left(\frac{\mathcal{E}_c}{\mathcal{E}_0} \right). \qquad (1.11)$$

Obviously, the smaller the value of $\mathcal{E}_0$ (i.e., the level of background numerical noise), the larger the critical predictable time T_c, because the noise-growth exponent κ is mainly determined by physical parameters of the chaotic system and its properties, and $\mathcal{E}_c$ is nearly of the order of magnitude of the true physical solution $\mathcal{P}$. This is why in CNS it is necessary to greatly decrease the background numerical noise $\mathcal{E}_0$ to a sufficiently tiny level.

Thus, CNS has a straightforward strategy that involves reducing *both* the truncation error and the round-off error as much as possible, so that the numerical simulation of chaos is guaranteed to be reliable and convergent (i.e., "clean") over a prescribed *sufficiently* long interval of time $t \in [0, T_c]$. In CNS, the truncation error is reduced to below a prescribed level by using a *sufficiently* high order Taylor expansion [33–36] in time and a spectral method

[16,61] in space with a fine enough spatial discretization. Moreover, the round-off error is reduced to below a specified threshold using multiple-precision (MP) floating-point arithmetic [62] with each number given to a *sufficiently* large number of significant digits. Detailed CNS algorithms are described in Chapter 2 (see also [38–42,53]) for temporal chaos and in Chapter 3 (see also [44,49–51]) for spatiotemporal chaos.

Truncation error is a well known concept in the field of computational mathematics, and affects all algorithms. Many studies have examined the truncation error of numerical algorithms used to solve different types of equations. In theory, it is not very difficult to reduce the truncation error to below a specified level. Taking (1.8) as an example, when $f(t + h)$ is approximated by

$$f(t) + \sum_{m=1}^{M} \frac{f^{(m)}(t)}{m!} h^m,$$

the higher the order M of the Taylor expansion, the smaller the truncation error R_T, provided the time-step h is sufficiently small (e.g. within the convergence radius). To date, the majority of numerical algorithms neglect round-off error, and, besides, multiple-precision (MP) floating-point arithmetic [62] is not widely implemented (even though MP has been used successfully to calculate π to an accuracy of millions of significant digits). In fact, multiple-precision (MP) floating-point arithmetic is freely available and easy to install on either a laptop or a supercomputer. The reader is recommended to visit https://gmplib.org/ to install the GMP (GNU multiple-precision floating-point arithmetic library), followed by https://mpfr.loria.fr/ to install the GNU MPRF library. The GMP website states, "arithmetic without limitations." This is almost true in that the round-off error can be reduced to *any* required level provided a *sufficiently* large number of significant digits is used to represent all real numbers.

Chaotic systems have hardly any perfectly exact solutions. In other words, true physical solutions are mostly unknown. Even so, it is self-evident that all numerical simulations u_i with negligible numerical noise should match the true physical solution $\mathcal{P}$. Hence,

$$\left| \mathcal{P} - u_i \right| < \epsilon/2, \quad t \in [0, T_c], \ i = 1, 2, 3, \cdots, N, \tag{1.12}$$

where $\epsilon > 0$ is a tiny constant and T_c is the critical predictable time. Then, we have

$$\left| u_i - u_j \right| \le \left| \mathcal{P} - u_i \right| + \left| \mathcal{P} - u_j \right| \le \epsilon, \quad t \in [0, T_c], \tag{1.13}$$

which is a *necessary* condition for a numerical simulation u_i to be close to its true physical solution $\mathcal{P}$. Moreover, owing to the butterfly effect of chaos and role of the exponential increment (1.10) in the numerical deviation, this must also be a *sufficient* condition. Such CNS results are regarded as "convergent" and therefore "reliable." In practice, the critical predictable time T_c

is determined by the *sufficient* and *necessary* condition,

$$|u_i - u_j| \leq \epsilon, \quad t \in [0, T_c], \tag{1.14}$$

obtained by comparing CNS results with those having even smaller background numerical noise.

CNS has the following objectives:

(a) To reduce background numerical noise to a specified (tiny) level.

(b) To determine the critical predictable time T_c.

(c) To effectively obtain a "clean" numerical simulation with a sufficiently large T_c.

(d) To use convergent, "clean" results over a specified long time interval $t \in [0, T_c]$ so as to deepen our understanding of chaos and turbulence, including investigation on their stability of statistics.

Although numerical algorithms based on high-order series approximations and multiple precision (MP) floating-point arithmetic have been utilised, to the best of the author's knowledge, the above key objectives have not yet been investigated systematically. It is the purpose of this book to address this knowledge gap.

The critical predictable time T_c plays a very important role in CNS. Unlike all other numerical methods, CNS focuses on the convergence and reliability of the trajectory of a chaotic system, especially its critical predictable time T_c, before which false numerical noise δ' is negligible compared to the true physical solution $\mathcal{P}$. At longer time $t > T_c$, a chaotic numerical simulation $\mathcal{S}$ quickly becomes a mixture of the true physical solution $\mathcal{P}$ and false numerical noise δ', both of which are of the same order of magnitude, say, $\delta' \sim \mathcal{P}$. In other words, a numerical simulation is "clean"[¶] before the critical predictable time T_c is reached, but is badly contaminated by numerical noise δ' afterwards. Unfortunately, unlike CNS, nearly all conventional algorithms of chaotic systems, including direct numerical simulation (DNS) [16, 17], overlook exponentially growing pollution by false numerical noise of chaotic systems and thus entirely neglect the concept of a "critical predictable time" T_c. The foregoing are *essential differences* between CNS and all other numerical methods including DNS [16, 17].

It is simply the case that the numerical simulation $\mathcal{S}'$ of a chaotic system produced by a traditional algorithm using single (or double) precision floating-point arithmetic is rapidly polluted by numerical noise δ', to the point that

---

¶Here, "clean" is relative in that false numerical noise δ' still exists, but is much smaller than the true physical solution $\mathcal{P}$, say, $|\delta'| \ll |\mathcal{P}|$, and thus negligible. This is like drinking water in a kitchen. Unlike pure water, drinking water might contain a few bacteria whose number is too small to be toxic.

false numerical noise δ' is of the same order of magnitude as the true physical solution $\mathcal{P}$, say, $\delta' \sim \mathcal{P}$. In theory, any statistical results based on such a badly polluted numerical simulation, i.e., the mixture $\mathcal{S}' = \mathcal{P} + \delta'$, can only be reliable, *if and only if* the statistical results are *not* inherently sensitive to small disturbances, say,

$$\langle \mathcal{P} + \delta' \rangle = \langle \mathcal{P} \rangle, \quad \text{when } \delta' \sim \mathcal{P} \text{ mostly,}$$

where $\langle \rangle$ denotes a statistical operator. Unfortunately, there is *no* serious proof for general cases and so this is merely a **hypothesis**! Many examples invalidating this hypothesis can be found using CNS, as illustrated in Chapter 5. In short, *not only* trajectories *but also* even the statistics of many chaotic systems are *unstable*, i.e., sensitive to small disturbances.

1.3 Some Illustrative Applications

For $t \in [0, T_c]$, the false numerical noise δ' of a CNS result is much lower than the true physical solution $\mathcal{P}$, say, $|\delta'| \ll |\mathcal{P}|$, and is thus negligible. This kind of "clean" result over a long time interval has hardly been reported previously, to the best of the author's knowledge. However, using the convergent trajectory of chaotic system gained by CNS as benchmark solution, many new, interesting results have been obtained. A few examples are briefly described below.

1.3.1 Convergent Chaotic Trajectory

In 2009, the author [38] successfully applied CNS to simulate a convergent chaotic trajectory of the Lorenz equations (1.6) and (1.7) over $t \in [0, 1000]$ by means of a 400th-order Taylor expansion using multiple-precision (MP) floating-point arithmetic with each real number represented by 800 significant digits. Convergence and reliability were confirmed by comparison against an additional simulation obtained by CNS using a 420th-order Taylor expansion and multiple-precision floating-point arithmetic with 820 significant digits. In 2014, the convergent chaotic trajectory of the Lorenz equations (1.6) and (1.7) was produced over a much longer time interval $t \in [0, 10000]$ by CNS using a 3500th-order Taylor expansion and multiple-precision floating-point arithmetic with 4180 digits, whose convergence and reliability were confirmed by an additional simulation of CNS using a 3600th-order Taylor expansion and multiple-precision floating-point arithmetic with 4515 digits [41]. It should be emphasized that chaotic numerical simulations of Lorenz equation given by the Runge-Kutta method in *double-precision* floating-point arithmetic are convergent over a rather smaller interval of approximately $t \in [0, 32]$ only,

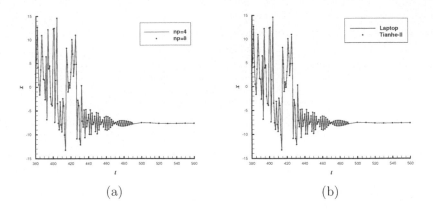

(a)   (b)

FIGURE 1.10
Numerical simulations of the trajectory $x(t)$ of Lorenz equations (1.6) for $\sigma = 10, b = 8/3, r = 23$, and initial value (1.7), obtained using a 200th-order Taylor's expansion method and multiple-precision floating-point arithmetic with each real number represented to 512 significant digits: (a) computations on the same laptop (Thinkpad L440 with Intel Core i7-4712MQ) but different number np of processors; and (b) computations using the same number of processors ($np = 4$) but on different computers, i.e., Thinkpad L440 laptop with Intel Core i7-4712MQ and supercomputer Tianhe-II.

such that $T_c \approx 32$, which is too short for practical purposes. Hence, by means of CNS, the critical predictable time T_c is increased by approximately three orders of magnitude for the Lorenz equations (1.6) and (1.7). Such a high order of Taylor series with multiple-precision floating-point arithmetic to so many significant digits had *not* been utilised previously, to the best of the author's knowledge. This illustrates that, in theory, CNS can indeed provide convergent numerical simulations of a chaotic trajectory over an *arbitrary* long (but finite) interval of time, provided sufficient computational resource is available. This also verifies the validity of the strategy behind CNS mentioned in Section 1.2. For details, please refer to [38, 41] and Section 2.3 of this book.

Recall the confused numerical simulations in Figures 1.6 to 1.9 obtained using various algorithms in *double-precision* floating-point arithmetic. As shown in Figure 1.10, the *same* convergent steady final value of $x_{1,1}(t)$ is obtained using the CNS algorithm based on a 200th-order Taylor's expansion and multiple-precision (MP) arithmetic with 512 significant digits, no matter whether computed on a laptop or a supercomputer or using four or eight processors, etc. This example provides an excellent demonstration that a CNS result can eliminate nearly all uncertainty (and confusion) introduced by conventional algorithms in double-precision floating-point arithmetic applied to a chaotic system. In other words, the CNS result is *not* sensitive to artificial disturbances, because it is very close to the true physical solution $\mathcal{P}$. Certainly, such kind of "clean" prediction by CNS can be used as a benchmark solution for comparisons.

Let us recall the pessimistic viewpoints reported in Section 1.1, namely, "*all* chaotic responses are simply numerical *noise* and have *nothing* to do with the solutions of differential equations" [12], and "reports of computed nonperiodic solutions of chaotic differential equations are simply consequences of unstably amplified truncation *errors*, and are not approximate solutions of the associated differential equations" [37]. We can now happily dispel these pessimistic viewpoints. Using CNS, we can obtain a *convergent, reliable* chaotic trajectory over a long enough time interval $t \in [0, T_c]$, during which false numerical noise δ' is relatively much smaller than the true physical solution $\mathcal{P}$, say, $|\delta'| \ll |\mathcal{P}|$, and therefore negligible! So, in the frame of the CNS, the intense debate about numerical simulation of chaos [12,13,37] can be settled. Thus, we have many reasons to believe that CNS should have an important role to play in the theory of chaos.

1.3.2 Evolution of Micro-level Physical Uncertainty

The background numerical noise of a reliable, convergent CNS result could be much smaller than even the micro-level physical uncertainty of a chaotic dynamic system. Thus, CNS provides us with a *new* way to *accurately* investigate the evolution and propagation of micro-level physical uncertainty in many chaotic systems.

For example, by means of CNS, it has been revealed by Liao and Li [42] that micro-level physical uncertainty in the initial condition of a chaotic three-body system leads invariably to macroscopic randomness, symmetry-breaking, and system disruption[||], without any external disturbances, i.e., seemingly out of *nothing*! In other words, the macroscopic randomness, symmetry-breaking and system disruption of this chaotic three-body system [8] appear to be *self-excited*. For further details, please refer to Section 4.1 of this book.

CNS has also been instrumental in providing rigorous theoretical evidence [44], for the first time, that micro-level physical thermal fluctuations should be the origin of the macroscopic randomness of Rayleigh-Bénard turbulent flows. For further details, please refer to Lin *et al.* [44] and Section 4.2 of this book.

1.3.3 Ultra-chaos and Statistic Instability

Due to the butterfly effect, the trajectories of a chaotic system are very sensitive to small disturbances. In other words, *all* chaotic systems exhibit *trajectory instability*. A numerical simulation S' of a chaotic trajectory given by

---

[||]Please see reference [8]

traditional algorithms in double-precision floating-point arithmetic is often badly polluted by numerical noise, and so is a mixture of the true physical solution $\mathcal{P}$ and false numerical noise δ', both of which are mostly of the same order of magnitude, say, $\delta' \sim \mathcal{P}$. Unfortunately, badly polluted numerical simulations (i.e., $\mathcal{S}' = \mathcal{P} + \delta'$) of chaos are widely used to calculate statistics $\langle \mathcal{S}' \rangle$ of chaotic systems, based on the assumption that the statistics $\langle \mathcal{S}' \rangle$ of chaos are *not* sensitive to numerical noise, say,

$$\langle \mathcal{P} + \delta' \rangle = \langle \mathcal{P} \rangle, \quad \text{when } \delta' \sim \mathcal{P} \text{ mostly,}$$

where $\langle \rangle$ denotes a statistical operator. But is this true? This is a very important, fundamental question.

In the frame of CNS, background numerical noise can be reduced to be much lower than physical and artificial small disturbances. So, CNS is capable of providing very "clean" solutions for benchmarking purposes so that the influence of small disturbances can be very accurately investigated. For example, the influence of a tiny level of external noise on the statistical properties of the spatiotemporal chaotic motion of a chain of pendulums coupled through elastic restoring and damped friction forces, governed by the damped, driven, sine-Gordon equation, has been carefully studied using CNS (with background numerical noise much lower than the tiny external noise). In this case, the simulation up to a critical predictable time T_c was gained by CNS [50], which is sufficiently long for statistical properties to be calculated (for details, see Section 5.2.1 of this book). Surprisingly, tiny external disturbances led to huge deviations in the statistical properties. It was found [49] that many chaotic systems have similar properties in that their statistics are sensitive to small disturbances, say,

$$\langle \mathcal{P} + \delta' \rangle \neq \langle \mathcal{P} \rangle, \quad \text{when } \delta' \sim \mathcal{P} \text{ mostly.}$$

It should be emphasized that it is CNS that made it possible to provide rigorous theoretical evidence [55] that numerical noise, as a kind of small-scale artificial stochastic disturbance, has both quantitative and qualitative large-scale influences on sustained turbulence (for details, see Chapter 6 of this book).

Two types of chaos exist: "normal-chaos", and "ultra-chaos", as classified by Liao and Qin [54]. The trajectories of both normal-chaos and ultra-chaos are sensitive to small disturbances. However, although the statistics of normal-chaos are *stable*, i.e., insensitive to small disturbances, the statistical properties such as the probability density function (PDF) of ultra-chaos are *unstable*, i.e., sensitive to tiny disturbances! It has been found that ultra-chaos widely exists, and thus has general scientific importance. It should be noted that the well-established term "hyper-chaos" defined by Rössler [63] in 1979 as a chaotic system with two positive Lyapunov exponents is likely to be a type of normal chaos, as demonstrated by Liao and Qin [54]. So, ultra-chaos is therefore a different concept to hyper-chaos. Several examples of ultra-chaos are given in Section 5.2 of this book.

As observed previously, a numerical simulation S' of chaos is often badly polluted by numerical noise, with false numerical noise δ' mostly of the same order of magnitude as the true physical solution $\mathcal{P}$, say, $\delta' \sim \mathcal{P}$. If such a system is ultra-chaotic, then its statistical results $\langle \mathcal{P}+\delta' \rangle$ based on such a kind of mixture (i.e., $S' = \mathcal{P} + \delta'$) are sensitive to artificial and/or environmental disturbances, and use of conventional algorithms (in single/double-precision floating-point arithmetic) would lead to different statistical results, say,

$$\langle \mathcal{P} + \delta' \rangle \neq \langle \mathcal{P} \rangle, \quad \text{when } \delta' \sim \mathcal{P} \text{ mostly,} \quad (1.15)$$

where $\langle \rangle$ denotes a statistical operator. This implies that the hypothesis (1.9) is valid for a normal-chaos only but *not* for an ultra-chaos. In other words, there exists not only **trajectory instability** but also **statistics instability** for an ultra-chaos!

Ultra-chaos has an important meaning. Statistical non-reproducibility is an inherent property of ultra-chaos, so that ultra-chaos is at a higher-level of disorder than normal-chaos. It is therefore impossible in practice to replicate experimental and/or numerical results of ultra-chaos, even in terms of statistics, because random environmental and/or artificial disturbances are ubiquitous and out of control. Thus, ultra-chaos might provide an insurmountable obstacle to reproducibility and repeatability. But, reproducibility and/or replicability form a cornerstone of modern science, confirming a principle of nature that scientific laws are invariant across space and time. Therefore, if ultra-chaotic systems do exist, the modern scientific paradigm becomes invalid due to the loss of reproducibility. So, the existence of ultra-chaos might lead to an *incompleteness* of modern scientific paradigm. This is indeed a very serious problem and a great challenge for us! To emphasize its importance, the author proposes the following two conjectures in Chapter 5 of this book:

> **Conjecture 5.1** *Statistics instability*: There should exist certain dynamical systems, whose statistics are unstable, i.e., sensitive to small disturbances. In other words, $\langle \mathcal{P}+\delta' \rangle$ is not always equal to $\langle \mathcal{P} \rangle$, where $\langle \rangle$ denotes a statistical operator, $S' = \mathcal{P} + \delta'$ is a numerical simulation of chaos obtained by conventional algorithms using single or double floating-point arithmetic, $\mathcal{P}$ denotes the true physical solution that can be accurately obtained by CNS, and δ' denotes false numerical noise, which all are mostly of the same order of magnitude, i.e., $\delta' \sim \mathcal{P} \sim S'$.

> **Conjecture 5.2** *Incompleteness of modern scientific paradigm*: There should exist some dynamic systems with statistics instability, for which the current paradigm of modern science is invalid due to the loss of reproducibility and/or replicability of physical/numerical experiments.

Hopefully some rigorous proofs (or disproofs) about these two conjectures will be developed in the near future. If true, a paradigm shift is needed for science.

1.3.4   Discovery of Periodic Orbits of Three-Body Problem

As mentioned in Section 1.1, the three-body problem concerning the motion of three-point masses under Newton's laws of gravitational attraction and motion, governed by (1.4), was first posed by Newton [18] in 1687. In the following 300 years, only three families of periodic orbits were found until 2013, when Šuvakov and Dmitrašinović [64] made a breakthrough discovery of 13 new distinct periodic orbits. Why is the three-body problem so difficult? The answer was revealed in 1890 by Poincaré [1], who pointed out that trajectories of three-body problem experience so-called "sensitivity dependence on initial condition" (SDIC), whereby a tiny variation in initial conditions can lead to an obvious difference in trajectories over a long interval of time. Hence, the motion of a three-body system is usually chaotic. To find a periodic orbit, it is therefore necessary to obtain sufficiently accurate trajectories over a sufficiently long time interval. This is very difficult to achieve using conventional algorithms in double-precision floating-point arithmetic, as mentioned in Section 1.1, but is straightforward using CNS, which can provide a reliable, convergent trajectory over a sufficiently long time interval $t \in [0, T_c]$, as illustrated in Section 1.3.1 and Section 2.3 of this book.

CNS has been used to discover more than six hundred new families of periodic orbits for a three-body system with *equal* masses, zero angular momentum, and initial conditions arranged in an isosceles collinear configuration [45]. Figure 1.11 shows some examples of newly discovered families of periodic three-body orbits. Besides, more than one thousand new families of periodic orbits of a three-body system with *unequal* masses have been discovered by Li, Jing and Liao [46] using CNS. These new families of periodic orbits formed the subject of the following articles in *New Scientist*:

1. Crane, L., "Infamous three-body problem has over a thousand new solutions", *New Scientist*, 20 September 2017 [65].

2. Whyte, C., "Watch the weird new solutions to the baffling three-body problem", *New Scientist*, 25 May 2018 [66].

Using CNS, the number of periodic orbits of the three-body problem has been increased by several orders of magnitude [52]. In addition, based on CNS in combination with machine learning, a roadmap to obtain periodic orbits of the three-body system with arbitrary masses has been suggested by Liao, Li and Yang [53]. For more details about these newly discovered periodic orbits, please refer to the references [45, 46, 52, 53] and Chapter 6 of this book, and visit the websites:

A. https://github.com/sjtu-liao/three-body,

B. https://numericaltank.sjtu.edu.cn/three-body/three-body.htm.

It should be emphasized that none of these periodic orbits had been reported previously. This again highlights the great potential of CNS, as a truly

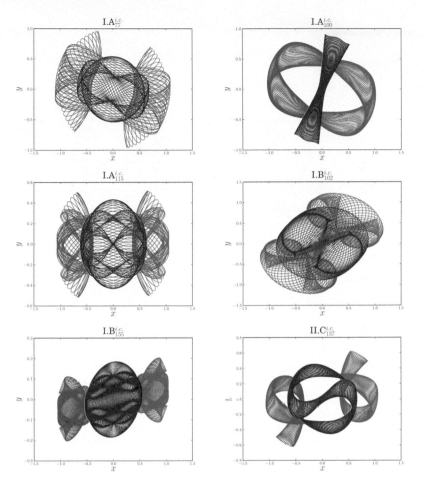

FIGURE 1.11

Periodic orbits of three-body system in case of equal mass, zero angular momentum and initial conditions with isosceles collinear configuration, discovered by Li & Liao [45] using the CNS-based strategy. Blue line: orbit of Body-1; red line, orbit of Body-2; black line, orbit of Body-3.

novel method, capable of bringing completely new insights to nonlinear physical systems. The three-body problem provides a good example to illustrate that CNS can be used to tackle such open problems. Indeed, CNS could lead to many new discoveries in different areas.

1.4   Outline

This book briefly presents the basic principles of Clean Numerical Simulation (CNS) and provides several illustrative applications. Chapter 2 systematically describes the fundamental concepts and numerical algorithms behind CNS for temporal chaos in systems with a finite number of degrees-of-freedom. Chapter 3 extends the concepts and algorithms to spatiotemporal chaos, which involves an infinite number of degrees-of-freedom. Chapters 2 and 3 address the following questions:

- How to reduce background numerical noise to below a required tiny level?

- How to determine the critical predictable time T_c?

- How to effectively obtain a "clean" numerical simulation with a large enough T_c?

Chapters 4 to 7 present results from some applications of CNS aimed at answering the following question:

- How to use "clean" results given by CNS, which are convergent and thus reliable over a sufficiently long time interval $t \in [0, T_c]$, to deepen our understanding of chaos and turbulence?

CNS is used to obtain convergent trajectories of chaos over a prescribed sufficiently long interval of time (Chapter 4), to investigate the origin of macroscopic randomness of some dynamic systems (Chapter 4), and to study the influence of tiny disturbances on the statistics instability of chaos, leading to the new concept of "ultra-chaos" (Chapter 5), and to investigate the influence of small levels of numerical noise on turbulence (Chapter 6), and to discover thousands of families of new orbits of Newton's three-body problem (Chapter 7). These applications serve to verify and validate CNS, while also illustrating its great potential for simulating problems involving chaos and turbulence.

In this book, the author proposes several conjectures (see Section 5.4) and open questions including the following *modified fourth Clay millennium problem* (see Section 6.6):

> **Open question 6.1** *The existence, smoothness and statistics stability of the Navier–Stokes equation*: Can we prove the existence and smoothness of the solution of the Navier–Stokes equation with physically proper boundary and initial conditions, whose statistics are stable (or unstable), say, insensitive (or sensitive) to small disturbances?

Although this open problem might be more difficult than the original fourth Clay millennium problem[**], the author firmly believes that statistics stability of turbulent flow is very important, especially from a practical viewpoint.

The book is written for use by final year undergraduate students and graduates with a basic knowledge of calculus, numerical methods, nonlinear dynamics, and chaos. Links (`https://github.com/sjtu-liao/CNS-code`) are given to freely available computer codes and animations of numerical simulations so that readers can apply CNS quickly and conveniently, leading to better understanding of its importance and potential. Detailed references are given for advanced readers.

It is worth restating the following. Numerical simulations S' of chaotic systems given by conventional algorithms in single (or double) precision floating-point arithmetic contain false numerical noise δ' that is mostly of the same order of magnitude as the true physical solution $\mathcal{P}$, say, $\delta' \sim \mathcal{P}$. In other words, a chaotic trajectory given by conventional algorithms is mostly badly polluted by numerical noise. On the contrary, a computer-generated chaotic simulation S given by CNS has much lower false numerical noise δ' in a long interval of time $t \in [0, T_c]$ than the true physical solution $\mathcal{P}$, i.e., $|\delta'| \ll |\mathcal{P}|$, thus the simulation S given by CNS is very close to the true physical solution $\mathcal{P}$, say, $S \approx \mathcal{P}$, so that one can gain $\langle \mathcal{P} \rangle$ accurately and then can check the statistics stability $\langle \mathcal{P} + \delta' \rangle = \langle \mathcal{P} \rangle$, where $\langle \rangle$ denotes a statistical operator, and $S' = \mathcal{P} + \delta'$ is an another simulation badly polluted by numerical noise δ' gained by conventional algorithms. In practice, the critical predictable time T_c can be increased by reducing the background numerical noise arising from round-off and truncation errors. Such kinds of "clean" computer-generated simulations of chaos over a prescribed sufficiently long interval of time have been *hardly* ever reported previously, and thus could lead to many completely *new* insights into nonlinear systems.

As a new, powerful tool for modelling chaotic systems and turbulence, CNS provides us with many opportunities and challenges. In short, CNS opens a new door, through which we can enter a "clean" numerical world!

[**]`http://www.claymath.org/millennium-problems`

2

CNS Algorithms for Temporal Chaos

As mentioned in the Introduction, Poincaré [1] was the first to discover, in 1890, the sensitivity dependence of a dynamic system to initial conditions (SDIC). Such sensitivity dependence demonstrated itself by a tiny variation in initial conditions leading to obvious deviation in the trajectory over a long time interval. SDIC was rediscovered by Lorenz [2] in 1963, and given the popular name "butterfly-effect" describing how a hurricane in North America might be created by the flapping of wings of a distant butterfly in South America several weeks earlier. The great discoveries by Poincaré [1] and Lorenz [2] ushered in a new research field called chaos dynamics [3,5,7,8,24], which is now regarded as one of the three greatest scientific revolutions in physics in the 20th century, comparable to quantum mechanics and Einstein's theory of relativity.

Lorenz [9,10] also discovered that computer-generated trajectories of chaos exhibited sensitivity dependence not only on initial conditions but also on artificial factors such as the numerical algorithm, time-step, and so on. SDIC is acceptable in physics where different initial conditions invariably have physical meaning. In computer science, however, the numerical algorithm, time-step, etc., are artefacts, and so sensitivity dependence on artificial factors (SDAF) has *no* physical meaning. Obviously, a convergent trajectory of a differential equation should have *nothing* to do with such artefacts.

Naturally, for chaotic systems, the occurrence of sensitivity dependence on artificial factors has led to much debate. For example, Yao and Hughes [12] came to the rather pessimistic conclusion that "all chaotic responses are *simply numerical noise* and have *nothing* to do with the solutions of differential equations." Obviously, the key point of the foregoing debate is whether or not one can obtain a *convergent* trajectory of chaos over a long enough interval of time. Theoretically speaking, this is indeed very important, because "it would be an exciting contribution if a convergent computed chaotic solution for a Lorenz model could be obtained" as pointed out by Yao and Hughes [12].

This chapter first describes the basic principles of Clean Numerical Simulation (CNS) [38–58] for temporal chaos. Using the famous Lorenz equation as an example, it is then illustrated that, from a mathematical perspective, *convergent* trajectories of chaotic systems over a prescribed sufficiently long interval of time are indeed *possible* by means of CNS. As pointed out by Yao and Hughes [12], this is indeed "an exciting contribution", which not only

DOI: 10.1201/9781003299622-2

settles the debate but also has very important repercussions for the theory of chaos.

2.1 Numerical Noise

When a continuous differential equation is solved numerically, truncation and round-off errors are inherent and unavoidable. Truncation error[*] is caused when an infinite series is limited to a finite number of terms. For example, in numerical simulations, the infinite series

$$e = 1 + \sum_{m=1}^{+\infty} \frac{1}{m!} \tag{2.1}$$

can be approximated by the *finite* terms

$$e \approx 1 + \sum_{m=1}^{M} \frac{1}{m!}, \tag{2.2}$$

with truncation error given by

$$R_T = \sum_{m=M+1}^{+\infty} \frac{1}{m!}. \tag{2.3}$$

Obviously, the larger the order M of the Taylor expansion (2.2), the smaller the corresponding truncation error R_T. In theory, the truncation error R_T given by (2.3) can be *arbitrarily* small, so long as the order M of the Taylor expansion (2.2) is sufficiently large, as listed in Table 2.1.

On the other hand, the round-off error[†] is due to the use of finitely many digits to represent real numbers that in theory have infinitely many digits and to carry out arithmetic operations with them, because only a finite amount of information may be stored at each stage of calculation process. Round-off error is always encountered in floating-point computations. In evaluating floating-point expressions, the magnitude of round-off error depends upon the hardware used: typically, double-precision floating-point arithmetic utilises 64 bits and is accurate to 16 decimal places. In practice, the round-off error can be easily reduced to below *any* prescribed arbitrarily small level by means of multiple-precision (MP) arithmetic [62], which uses a *sufficiently* large number of significant digits to represent floating-point data. The multiple-precision

---

[*]https://en.wikipedia.org/wiki/Truncation_error
[†]https://en.wikipedia.org/wiki/Round-off_error

TABLE 2.1

The Truncation Error R_T of the Taylor Expansion (2.2) at the Order M

M	R_T
5	1.62×10^{-3}
10	2.73×10^{-8}
15	5.08×10^{-14}
20	2.05×10^{-20}
30	1.26×10^{-34}
40	3.06×10^{-50}
50	6.57×10^{-67}
100	1.07×10^{-160}

software is easy (and free) to install. For details, please visit the website https://gmplib.org/ to install the GMP (GNU multiple precision arithmetic library) and then https://mpfr.loria.fr/ to install the GNU MPRF library. The GMP website contains the following words: "arithmetic without limitations."

As pointed out by Parker and Chua [24], local round-off and truncation errors propagate together in a rather complicated way, leading to so-called global round-off and global truncation errors. Background numerical noise is determined as the maximum of the global truncation error and global round-off error. Obviously, when the truncation error is much smaller than the round-off error, the background numerical noise is determined by the round-off error. Thus, when double precision floating-point arithmetic is used, the background numerical noise stops decreasing when the truncation error is smaller than the round-off error so that one *cannot* further reduce background numerical noise to below a required arbitrarily small level. This is exactly the reason why one cannot obtain convergent chaotic simulations by means of double precision floating-point arithmetic [9–12, 14, 32, 37]. Therefore, it is *useless* to reduce either the truncation error or the round-off error *in isolation*. In other words, we must maintain a synchronized decrease in *both* the truncation error and round-off error so as to reduce the background numerical noise to a required *arbitrarily* small level! This is the key strategy underpinning CNS.

It should be emphasized that *all* the aforementioned authors [9–12, 14, 32, 37], whose viewpoints about numerical simulations of chaos were rather negative and pessimistic, used *double precision* floating-point arithmetic in the numerical models described in their publications. It is indeed unfortunate that the majority of researchers have neglected the influence of round-off error on chaotic systems, inevitably leading to confusion and stoking the heated debates mentioned earlier [11, 12].

2.2 Basic Principles of CNS for Temporal Chaos

Let us consider a dynamic system of temporal chaos, governed by the following set of ordinary differential equations (ODEs)

$$\dot{u}_i = f_i(\mathbf{u}, t), \qquad 1 \le i \le N, \tag{2.4}$$

where t denotes time, the dot denotes differentiation with respect to t, N is an integer, and

$$\mathbf{u} = \left(u_1, u_2, u_3, \cdots, u_N\right)$$

is a vector of the unknown function $u_i(t)$ for $1 \le i \le N$. The initial condition at $t = 0$ reads

$$\mathbf{u}(0) = \mathbf{u}_0 = \left(\alpha_1, \alpha_2, \alpha_3, \cdots, \alpha_N\right), \tag{2.5}$$

where α_i is a given constant for $1 \le i \le N$.

Conventionally, Eqs. (2.4) and (2.5) could be numerically solved by means of the fourth-order Runge-Kutta method in double-precision floating-point arithmetic. However this *cannot* provide a convergent chaotic trajectory over a prescribed sufficiently long time interval [9–12, 14, 32, 37].

2.2.1 How to Reduce Background Numerical Noise

The background numerical noise $\mathcal{E}_0$ is determined as the maximum of the truncation error and round-off error. In the frame of CNS, to reduce the truncation error to below a required arbitrarily small level, we use a high-order Taylor expansion

$$u_i(t + \Delta t) \approx u_i(t) + \sum_{m=1}^{M} u_i^{[m]}(t)(\Delta t)^m, \tag{2.6}$$

where M is the order of the Taylor expansion, Δt is the time-step, and

$$u_i^{[m]}(t) = \frac{1}{m!} \frac{d^m u_i(t)}{dt^m}, \qquad 1 \le i \le N. \tag{2.7}$$

Differentiating $(m-1)$ times both sides of the governing equation (2.4) with respect to time t and then dividing by $m!$, we arrive at the explicit expression

$$u_i^{[m]}(t) = \frac{1}{m!} \frac{d^{m-1} f_i(\mathbf{u}, t)}{dt^{m-1}}, \qquad 1 \le i \le N. \tag{2.8}$$

Obviously, the higher the order M of the Taylor expansion (2.6), the smaller its truncation error, so long as the time-step Δt remains less than its radius of convergence. In this way, we can reduce the truncation error to below a required arbitrarily small level.

Moreover, multiple-precision (MP) floating-point arithmetic with N_s significant digits [62] instead of double precision floating-point arithmetic is used for *all* floating-point data. Obviously, the round-off error can be reduced to below a required arbitrarily small level, provided the number N_s of the significant digits is sufficiently large. In this way, *both* the truncation error and the round-off error can be reduced to below a prescribed *arbitrarily* small level.

2.2.2   Determination of Critical Predictable Time T_c

It is well known that two nearby trajectories of a temporal chaos exponentially deviate from each other [5, 7, 8, 24, 25]. This process is characterized by the leading positive Lyapunov exponent. Accordingly, the false numerical noise also increases exponentially, such that

$$\mathcal{E}(t) = \mathcal{E}_0 \exp(\kappa\, t), \tag{2.9}$$

where $\kappa > 0$ is the so-called "noise-growth exponent" that is normally equal to the leading positive Lyapunov exponent for temporal chaos, $\mathcal{E}_0$ denotes the level of the background numerical noise (determined as the maximum of the truncation error and the round-off error), $\mathcal{E}(t)$ is the level (on average) of simulation deviation from the true physical solution $\mathcal{P}$.

Therefore, a computer-generated simulation $\mathcal{S}$ of a chaotic system is a mixture of the true physical solution $\mathcal{P}$ and false numerical noise δ', say, $\mathcal{S} = \mathcal{P} + \delta'$. Initially, the false numerical noise δ' is much smaller than the true physical solution $\mathcal{P}$, i.e., $|\delta'| \ll |\mathcal{P}|$, causing the numerical noise δ' to be negligible; in which case, the computer-generated numerical simulation $\mathcal{S}$ can be regarded as "convergent" and then "reliable", with $\mathcal{S} \approx \mathcal{P}$. However, for a chaotic system, the false numerical noise δ' increases exponentially so that its magnitude quickly enlarges to reach the *same* order of the true physical solution $\mathcal{P}$, say, $\delta' \sim \mathcal{P}$. Thus, there exists a critical time T_c such that, within $t \in [0, T_c]$, the numerical noise δ' is negligible compared to the true physical solution $\mathcal{P}$, i.e., $|\delta'| \ll |\mathcal{P}|$, and as a result, the numerical simulation $\mathcal{S}$ of a trajectory is a sufficiently accurate representation of the true physical solution $\mathcal{P}$, say, $\mathcal{S} \approx \mathcal{P}$, and thus is reliable. However, beyond T_c, the false numerical noise δ' might attain the same order of magnitude as the true physical solution $\mathcal{P}$, say, $\delta' \sim \mathcal{P}$, causing the computer-generated simulation $\mathcal{S}$ to become unreliable. We call T_c "the critical predictable time", which is one of most important concepts in the frame of CNS.

Let $\mathcal{E}_c$ denote a critical level of numerical noise, which is nearly of the same order of magnitude as the true physical solution $\mathcal{P}$. Then, according to the earlier definition of the critical predictable time T_c, we have

$$\mathcal{E}(t) = \mathcal{E}_0 \exp(\kappa\, t), \qquad t \in [0, T_c], \tag{2.10}$$

and

$$\mathcal{E}_c = \mathcal{E}_0 \exp(\kappa\, T_c), \tag{2.11}$$

which gives

$$T_c = \frac{1}{\kappa} \ln \left(\frac{\mathcal{E}_c}{\mathcal{E}_0} \right), \tag{2.12}$$

where $\mathcal{E}_0$ is the background numerical noise and $\kappa > 0$ is the so-called "noise-growth exponent", respectively. Note that the critical numerical noise $\mathcal{E}_c$ is mainly determined by the order of magnitude of the true physical solution $\mathcal{P}$. So, for a given $\mathcal{E}_c$, the smaller the background numerical noise $\mathcal{E}_0$, the larger the critical predictable time T_c. This is the reason why we have to reduce the background numerical noise $\mathcal{E}_0$ to below the required small level.

For temporal chaos, $\kappa > 0$ is often equal to the leading positive Lyapunov exponent. In this case, one can determine T_c by means of (2.12). However, it is hard to determine exactly the background numerical noise $\mathcal{E}_0$. In practice, one can determine T_c for a chaotic trajectory by comparing the trajectory with a different trajectory obtained using the same equation for even smaller background numerical noise: the decoupling time of the two trajectories equates to the critical predictable time T_c of the former trajectory, as suggested by Teixeira *et al.* [11] and confirmed by Liao [38].

It is found by Teixeira *et al.* [11] that, for a given numerical algorithm, the critical predictable time T_c of chaotic trajectories given by different time-steps Δt generally follows a logarithmic rule, say,

$$T_c \approx A_0^R + A_1^R \ln (\Delta t), \tag{2.13}$$

where A_0^R and A_1^R are constants. Thus, the time-step Δt should be exponentially small for a given critical predictable time T_c; however, this is impossible in practice.

Besides, Liao [38] found that, for temporal chaos associated with a *fixed* time-step Δt, it generally holds that

$$T_c \approx a_1^R N_s + a_0^R, \tag{2.14}$$

when the truncation error is smaller than the round-off error, and

$$T_c \approx b_1^R M + b_0^R, \tag{2.15}$$

when the round-off error is smaller than the truncation error, where N_s is the number of significant digits of the multiple-precision (MP) floating-point arithmetic, M is the order of the Taylor expansion in (2.6), and $a_0^R, a_1^R, b_0^R, b_1^R$ are constants. Here, the values of the constants A_0^R, A_1^R in (2.13), a_0^R, a_1^R in (2.14) and b_0^R, b_1^R in (2.15) are determined by linear regressions of a few test simulations. Thus, for a fixed time-step Δt, we have

$$T_c \approx \frac{1}{\gamma} \min \left\{ a_1^R N_s + a_0^R, b_1^R M + b_0^R \right\}, \tag{2.16}$$

where $\gamma \geq 1$ is a constant, which is used here as a kind of safety factor. Note that the regression coefficients b_0^R and b_1^R are dependent upon the time-step

Δt. However, a_0^R and a_1^R are independent upon Δt, because (2.14) is satisfied when the truncation error is smaller than the round-off error. For details, please refer to Liao [38–40].

2.2.3 Balance between Truncation and Round-off Error

Note that the background numerical noise $\mathcal{E}_0$ is determined by the maximum of the truncation error and round-off error. So, in order to save on computer resources and to decrease the required CPU time, it is better to keep the round-off error approximately at the same level as the truncation error. For this purpose, given a value of T_c and a *fixed* time-step Δt, we can obtain, using (2.14) and (2.15), the required number of significant digits of the multiple-precision (MP) arithmetic from

$$N_s \approx \left\lceil \frac{\gamma \, T_c - a_0^R}{a_1^R} \right\rceil \tag{2.17}$$

and the required order of Taylor expansion

$$M \approx \left\lceil \frac{\gamma \, T_c - b_0^R}{b_1^R} \right\rceil, \tag{2.18}$$

where $\lceil x \rceil$ is an operator that takes the integer part of x, and $\gamma \geq 1$ is used here as a kind of safety factor.

Alternatively, one can apply a *variable* time-step (VS) scheme [67] within the aforementioned CNS algorithm with a given allowable tolerance *tol* of the governing equation (2.4). According to Barrio *et al.* [67], the optimal time-step is given by

$$\Delta t = \min \left(\frac{tol^{\frac{1}{M}}}{\left\| u_i^{[M-1]}(t) \right\|_\infty^{\frac{1}{M-1}}}, \frac{tol^{\frac{1}{M+1}}}{\left\| u_i^{[M]}(t) \right\|_\infty^{\frac{1}{M}}} \right), \qquad 1 \leq i \leq N, \tag{2.19}$$

where M denotes the order of Taylor expansion (2.6), *tol* denotes the allowable tolerance of (2.4), $\left\| u_i^{[M]}(t) \right\|_\infty$ is the infinite norm of the variable $u_i^{[M]}(t)$ ($1 \leq i \leq N$), and $u_i^{[M]}(t)$ is defined by (2.7), respectively. In this case, to achieve high computational efficiency, one can use the empirical formula

$$M = -c_1^E \log_{10}(tol) + c_0^E \tag{2.20}$$

to determine a proper order of Taylor expansion [67], where c_0^E and c_1^E are constants.

Furthermore, given that the round-off error should be of the same level as the truncation error, we should have

$$tol = 10^{-N_s}, \tag{2.21}$$

where N_s denotes the number of significant digits required in the multiple precision (MP) arithmetic. Substituting (2.21) into (2.19) and (2.20) gives the optimal time-step

$$\Delta t = \min\left(\frac{10^{-\frac{N_s}{M}}}{\left\|u_i^{[M-1]}(t)\right\|_\infty^{\frac{1}{M-1}}}, \frac{10^{-\frac{N_s}{M+1}}}{\left\|u_i^{[M]}(t)\right\|_\infty^{\frac{1}{M}}}\right), \qquad 1 \le i \le N, \qquad (2.22)$$

and the empirical relationship

$$M = c_1^E N_s + c_0^E. \qquad (2.23)$$

Jorba & Zou [68] suggested an optimal order of Taylor expansion

$$M = \left\lceil -\frac{1}{2}\ln(tol) + 1 \right\rceil, \qquad (2.24)$$

where *tol* is the allowable tolerance, and $\lceil x \rceil$ is an operator that takes the integer value of x. Substituting (2.21) into the above expression gives the optimal order of the Taylor expansion method:

$$M = \left\lceil \left(\frac{\ln 10}{2}\right) N_s + 1 \right\rceil \approx \left\lceil 1.15 N_s + 1 \right\rceil. \qquad (2.25)$$

As pointed out by Jorba & Zou [68], the above expression works under certain conditions. According to the author's experience, a constant coefficient larger than 1.15 should be used in general when parallel computation is applied.

In this way, we can control the background numerical noise $\mathcal{E}_0$ by selecting a prescribed number N_s of significant digits in multiple-precision arithmetic and ensuring a balance between the truncation error and round-off error via (2.20) and (2.21) with an optimal time-step specified by (2.19). For further details, please refer to [67–70].

Note that one key point of CNS is to determine the critical predictable time T_c. A balance should exist between the truncation error and round-off error, and so it is reasonable to assume that the background numerical noise should be equal to the round-off error, such that $\mathcal{E}_0 = 10^{-N_s}$, where N_s is the number of significant digits used in the multiple-precision (MP) arithmetic. Then, according to (2.10), we have

$$\mathcal{E}(t) = 10^{-N_s}\exp(\kappa t) = 10^{-N_s + \kappa t/\ln 10}, \qquad t \in [0, T_c], \qquad (2.26)$$

where $\kappa > 0$ is the noise-growth exponent. Furthermore,

$$\mathcal{E}_c = 10^{-N_s + \kappa T_c/\ln 10}, \qquad (2.27)$$

which gives the relationship between the number of significant digits N_s in the multiple-precision (MP) arithmetic and the critical predictable time T_c as:

$$N_s \approx \left\lceil \frac{\gamma\,\kappa\,T_c}{\ln 10} - \log_{10}\mathcal{E}_c \right\rceil, \qquad (2.28)$$

where $\mathcal{E}_c$ denotes the critical numerical noise (which is close to the order of magnitude of the true physical solution $\mathcal{P}$), $\kappa > 0$ is the positive leading Lyapunov exponent, $\lceil x \rceil$ is an operator that takes the integer value of x, and $\gamma \geq 1$ is a constant used here as a kind of safety factor. Then, using (2.23), we obtain the following relationship between the order M of the Taylor expansion and the critical predictable time T_c:

$$M \approx \left\lceil c_1^E \left(\frac{\gamma \kappa T_c}{\ln 10} - \log_{10} \mathcal{E}_c \right) + c_0^E \right\rceil. \tag{2.29}$$

According to (2.17) and (2.28), the number of significant digits N_s of the multiple-precision (MP) arithmetic should increase *linearly* with respect to the critical predictable time T_c. Besides, according to (2.18) and (2.29), the order M of Taylor expansion should increase *linearly* with respect to the critical predictable time T_c, too. The above conclusions have general meaning [38, 42]. They explain why one cannot gain convergent chaotic simulations over a sufficiently long time interval by means of conventional algorithms in double precision [9–12, 14, 32, 37], even if Taylor expansions at rather high orders are used. Unfortunately, it also means that it is expensive to gain a convergent trajectory of chaos over a prescribed long time interval.

From the author's experience, the critical predictable time T_c always increases linearly with respect to the number of significant digits N_s utilised in the multiple-precision arithmetic. As yet there is no general mathematical proof available, and so the following conjecture is proposed:

Conjecture 2.1 *Conjecture of T_c*: the critical predictable time T_c of a numerical simulation of a fully chaotic system should increase linearly with respect to the number of significant digits N_s used in the multiple-precision representation of all floating-point data, provided the truncation error is not larger than the round-off error.

Eq. (2.28) implies that the required number of significant digits N_s should linearly increase with respect to the critical predictable time T_c and the noise-growth exponent κ. This leads to a second conjecture:

Conjecture 2.2 *Conjecture of N_s*: for a fully chaotic dynamic system, the required number of significant digits N_s in the multiple-precision representation of all floating-point data should increase linearly with respect to the critical predictable time T_c and the noise-growth exponent κ, provided the truncation error is not larger than the round-off error.

2.2.4   Self-adaptive CNS Algorithm with Variable Precision

The background numerical noise $\mathcal{E}_0$ is determined by the maximum of the truncation error and round-off error. According to (2.10), it is necessary to use a very small level of background numerical noise $\mathcal{E}_0$ to gain a convergent chaotic simulation over a prescribed, sufficiently long time interval. This is indeed true. For example, in order to gain the convergent trajectory of the chaotic Lorenz system in $t \in [0, 10000]$, Liao & Wang [41] used a 3500th-order Taylor expansion ($M = 3500$) with time-step $\Delta t = 0.01$ in 4180-digit multiple precision ($N_s = 4180$). The corresponding background numerical noise was indeed very small. But, unfortunately, the computations were rather time-consuming.

According to (2.10), the numerical noise $\mathcal{E}(t)$ increases exponentially. Thus, after a time duration t^*, $\mathcal{E}$ becomes much larger than the background numerical noise $\mathcal{E}_0$, say, $\mathcal{E}(t^*) \gg \mathcal{E}_0$. So, it is *unnecessary* to keep the background numerical noise $\mathcal{E}_0$ *fixed* throughout the *whole* interval $t \in [0, T_c]$. In theory, according to (2.10), use of a larger background numerical noise $\mathcal{E}_0^* > \mathcal{E}_0$ when $t > t^*$ does not influence the CNS result, so long as $\mathcal{E}_0^* \leq \mathcal{E}(t^*)$.

Note that a larger round-off error corresponds to a multiple-precision (MP) with a smaller number N_s of significant digits, and a larger truncation error corresponds to a smaller order M of Taylor expansion, or a larger allowable tolerance $tol = 10^{-N_s}$ that leads to a *smaller* order

$$M = \left\lceil -c_1^E \log_{10}(tol) + c_0^E \right\rceil$$

of Taylor expansion. For CNS algorithms with a *fixed* time-step, according to (2.17) and (2.18), it is sufficient after integrating a time interval $t^* < T_c$ to use a smaller number of significant digits

$$N_s \approx \left\lceil \left| \frac{\gamma(T_c - t^*) - a_0^R}{a_1^R} \right| \right\rceil, \qquad (2.30)$$

and a lower order of Taylor expansion

$$M \approx \left\lceil \left| \frac{\gamma(T_c - t^*) - b_0^R}{b_1^R} \right| \right\rceil, \qquad (2.31)$$

to obtain the CNS result for $t \geq t^*$. In practice, it is unnecessary to change N_s and M at each time-step but instead at certain given times, such as $t^* = n\Delta T$ where $n \geq 1$ is an integer and $\Delta T = 25, 50, 100$, etc.

Similarly, for CNS algorithms with a *variable* time-step, after a simulation duration t^*, Eqs. (2.28) and (2.29) imply a smaller significant digit number,

$$N_s = \left\lceil \frac{\gamma\kappa(T_c - t^*)}{\ln 10} - \log_{10} \mathcal{E}_c \right\rceil, \qquad (2.32)$$

and a lower order of Taylor expansion,

$$M = \left\lceil c_1^E \left(\frac{\gamma\kappa(T_c - t^*)}{\ln 10} - \log_{10} \mathcal{E}_c \right) + c_0^E \right\rceil, \qquad (2.33)$$

where $\lceil x \rceil$ denotes an operator that takes the integer value of x, and $\gamma \geq 1$ is a positive constant used here as a kind of safety factor.

In this way, one can greatly increase the computational efficiency for a given critical predictable time T_c, as illustrated later. This kind of self-adaptive CNS algorithm with variable multiple-precision (MP) arithmetic leads to much higher computational efficiency than the previous CNS with fixed background numerical noise, as shown below.

2.3   An Illustrative Example: Lorenz Model

Let us consider here the famous Lorenz equation [2]:

$$\begin{cases} \dot{x} = -\sigma x + \sigma y, \\ \dot{y} = rx - y - x\,z, \\ \dot{z} = x\,y - bz, \end{cases} \tag{2.34}$$

subject to the initial condition

$$x(0) = -15,8, \quad y(0) = -17.48, \quad z(0) = 35.64, \tag{2.35}$$

in the chaotic case of $\sigma = 10$, $b = 8/3$, and $r = 28$, where $x(t), y(t)$, and $z(t)$ are unknown variables, and the dot denotes differentiation with respect to the time t.

In the frame of CNS, to reduce the truncation error to below a required arbitrarily small level, we use the Mth-order Taylor expansion

$$\begin{cases} x(t + \Delta t) = \sum\limits_{m=0}^{M} x^{[m]}(t)\,(\Delta t)^m, \\ y(t + \Delta t) = \sum\limits_{m=0}^{M} y^{[m]}(t)\,(\Delta t)^m, \\ z(t + \Delta t) = \sum\limits_{m=0}^{M} z^{[m]}(t)\,(\Delta t)^m, \end{cases} \tag{2.36}$$

where

$$x^{[m]}(t) = \frac{1}{m!}\frac{d^m x(t)}{dt^m}, \quad y^{[m]}(t) = \frac{1}{m!}\frac{d^m y(t)}{dt^m}, \quad z^{[m]}(t) = \frac{1}{m!}\frac{d^m z(t)}{dt^m},$$

with the definitions

$$x^{[0]}(t) = x(t), y^{[0]}(t) = y(t), z^{[0]}(t) = z(t).$$

Differentiating $(m-1)$ times $(m \geq 2)$ both sides of Eqs. (2.34) with respect to t and then dividing by $m!$, we have

$$
\begin{cases}
x^{[m]} = \frac{\sigma}{m}\left(-x^{[m-1]} + y^{[m-1]}\right), \\
y^{[m]} = \frac{1}{m}\left(r\, x^{[m-1]} - y^{[m-1]} - \sum_{j=0}^{m-1} x^{[j]}\, z^{[m-1-j]}\right), \\
z^{[m]} = \frac{1}{m}\left(\sum_{j=0}^{m-1} x^{[j]}\, y^{[m-1-j]} - b\, z^{[m-1]}\right).
\end{cases}
\tag{2.37}
$$

Besides, in order to reduce the round-off error to below a required arbitrarily small level, all floating-point data are represented in multiple-precision arithmetic with sufficient significant digits N_s. Note that the background numerical noise is determined by the maximum of the truncation error and round-off error. So, both the truncation error and the round-off error must be reduced synchronously.

2.3.1 Fixed Time-Step

Using a fixed time-step $\Delta t = 0.01$, Liao [38] discovered the following regression formulas

$$
T_c \approx 3M,
\tag{2.38}
$$

when the number of significant digits N_s is sufficiently large, and

$$
T_c \approx 2.51 N_s - 4.26,
\tag{2.39}
$$

when the order M of the temporal Taylor expansion is sufficiently large. According to (2.17) and (2.18), the corresponding regression constants are $a_0^R = -4.26$, $a_1^R = 2.51$, and $b_0^R = 0$, $b_1^R = 3$, respectively.

By means of a parallel CNS algorithm with a 400th-order Taylor expansion (i.e., $M = 400$), 800 significant digit multiple-precision arithmetic (i.e., $N_s = 800$), and a fixed time-step $\Delta t = 0.01$, Liao [38] obtained, *for the first time*, a convergent chaotic trajectory of the Lorenz equation (2.34) for the initial condition (2.35) over the time interval $t \in [0, 1000]$. Similarly, using a 1000th-order Taylor expansion (i.e., $M = 1000$), 2100 significant digit multiple-precision arithmetic (i.e., $N_s = 2100$), and a fixed time-step $\Delta t = 0.01$, Wang *et al.* [71] obtained a convergent chaotic trajectory over a longer time interval $t \in [0, 2500]$ by means of CNS. Furthermore, using a 3500th-order Taylor series (i.e., $M = 3500$), 4180 significant digit multiple-precision arithmetic (i.e., $N_s = 4180$), and a fixed time-step $\Delta t = 0.01$, Liao and Wang [41] successfully gained a convergent chaotic trajectory of the Lorenz equation over a very long interval of time $t \in [0, 10000]$, requiring 220.9 hours and 1200 CPUs of the National Supercomputer TH-1A, Tianjin, China. Its convergence and reliability were verified by means of the same CNS algorithm using a 3600th-order Taylor expansion (i.e., $M = 3600$) and 4515 significant digit multiple-precision arithmetic (i.e., $N_s = 4515$), whereby the even larger

TABLE 2.2
Convergent Chaotic Trajectory of Lorenz Equation (2.34) and (2.35)
Obtained for $\sigma = 10$, $r = 28$, and $b = -8/3$ Using a CNS Parallel Algorithm
with a 3500th-Order Taylor Expansion ($M = 3500$), 4180 Significant Digit
($N_s = 4180$) Multiple-Precision Arithmetic, and Time-Step $\Delta t = 0.01$,
Computed on 1200 CPUs of the National Supercomputer TH-1A, Tianjin,
China

t	$x(t)$	$y(t)$	$z(t)$
500	−5.3050	−9.4260	12.3022
1000	13.8820	19.9183	26.9019
1500	−10.1398	−7.6264	31.8584
2000	−6.8739	−1.4848	31.3495
2500	2.7592	0.4763	24.6411
3000	1.6933	3.6003	21.4109
3500	0.7357	−2.1187	24.4667
4000	−7.6927	−13.4996	14.1994
4500	−13.7455	−8.3158	38.8589
5000	−6.0844	−10.8137	12.7391
5500	4.7719	8.8154	10.4386
6000	0.2167	2.1043	22.1246
6500	4.6758	5.6919	20.4906
7000	−11.3949	−16.5754	23.6813
7500	0.1858	0.6489	16.5550
8000	−1.2659	−2.3363	17.4960
8500	−3.0412	1.5314	27.8442
9000	13.4797	17.2821	29.2382
9500	8.9996	3.0374	33.8242
10000	−15.8173	−17.3669	35.5584

M and N_s correspond to smaller truncation error and round-off error. Table 2.2 lists the convergent chaotic trajectory of the Lorenz equation over the interval $t \in [0, 10000]$. It should be emphasized that the convergent chaotic trajectory in such a long interval of time had *not* been reported previously! This provides evidence, *for the first time*, that it is possible to gain a convergent trajectory of chaotic systems over a prescribed long time interval. As pointed out by Yao and Hughes [12], this is indeed "an exciting contribution", which settles the heated debate concerning the convergence and reliability of numerical simulations of chaotic trajectories.

To improve computational efficiency, a self-adaptive CNS algorithm with *variable* multiple-precision arithmetic is recommended. For a *fixed* time-step of $\Delta t = 0.01$, according to (2.30) and (2.31), one could decrease the number of significant digits N_s in multiple-precision and the order M of the Taylor expansion for the time interval $t^* \leq t \leq T_c$ in the following way:

$$N_s = \left\lceil \frac{\gamma(T_c - t^*) + 4.26}{2.51} \right\rceil \approx \left\lceil \frac{\gamma(T_c - t^*)}{2.51} \right\rceil, \quad \text{when } T_c - t^* \geq 500, \quad (2.40)$$

and

$$M \approx \left\lceil \frac{\gamma(T_c - t^*)}{3} \right\rceil, \qquad \text{when } T_c - t^* \geq 500, \qquad (2.41)$$

where $\gamma \geq 1$ is used here as a safety factor, $t^* = n\Delta T$ with $n \geq 1$ an integer, and $\Delta T > 0$ a constant. For $\gamma = 1.1$ and $\Delta T = 100$, it took 95.6 hours to gain the *same* convergent chaotic solution of the Lorenz equation in $t \in [0, 10000]$ by means of the self-adaptive CNS algorithm with adjustable multiple-precision (2.40) using 1200 CPUs of the National Supercomputer Tianhe-II, Guangzhou, China. Compared to the 220.9 hours used by Liao and Wang [41] in 2014, the self-adaptive CNS algorithm with fixed time-step needs only 44.3% of the CPU time to obtain the *same* chaotic trajectory in $t \in [0, 10000]$.

2.3.2   Variable Optimal Time-step

At each time-step, one can use a variable optimal time-step given by (2.22), where the empirical relationship

$$M = \left\lceil 1.5\, N_s \right\rceil$$

is applied for the Lorenz model, together with a self-adaptive number of significant digits N_s in the multiple-precision arithmetic determined by (2.32). For the same chaotic case under consideration, the corresponding maximum Lyapunov exponent is equal to 0.91, i.e., $\kappa = 0.91$. Substituting this value into (2.32) and choosing $\varepsilon_c = 10^{-2}$, we obtain the following relationship

$$N_s = \left\lceil \frac{\gamma(T_c - t^*)}{2.53} + 2 \right\rceil \approx \left\lceil \frac{\gamma(T_c - t^*)}{2.53} \right\rceil, \qquad (2.42)$$

which closely agrees with the regression expression (2.40), where $\gamma \geq 1$ is used here as a safety factor. This verifies that the relationship (2.32) between N_s and T_c indeed has general meaning.

For $\gamma = 1.1$, $\Delta T = 100$ and $T_c - t^* \geq 500$, the self-adaptive CNS algorithm with adjustable multiple-precision (2.42) takes 37.4 hours to obtain the *same* chaotic trajectory of the Lorenz equation in $t \in [0, 10000]$ using the 1200 CPUs of the National Supercomputer Tianhe-II at Guangzhou, China. Compared to the 220.9 hours required for the CNS result by Liao and Wang [41] in 2014, the self-adaptive CNS algorithm with a variable optimal time-step needed only 16.9% of the CPU time to obtain the *same* chaotic trajectory in the same time interval $t \in [0, 10000]$. This illustrates and also verifies that a self-adaptive CNS algorithm with a variable optimal time-step can indeed greatly increase the computational efficiency of CNS. For more details, please refer to Qin and Liao [58]. The corresponding CNS codes for Lorenz model can be downloaded via the website `https://github.com/sjtu-liao/CNS-code`.

2.4 Significance of Convergent Chaotic Trajectory

It should be emphasized that the *convergent* chaotic trajectory of the Lorenz model over such long time intervals has *not* been reported previously! Let us again recall the earlier pessimistic comments that "*all* chaotic responses are simply numerical *noise* and have *nothing* to do with the solutions of differential equations" [12], and "reports of computed non-periodic solutions of chaotic differential equations are simply consequences of unstably amplified truncation *errors*, and are not approximate solutions of the associated differential equations" [37]. As pointed out by Yao and Hughes [12], it is indeed "an *exciting* contribution" to obtain this convergent chaotic trajectory over such a long interval of time, and so thoroughly settle the issues [9–14, 32, 37].

Obviously, all CNS results lie close to the true physical solution $\mathcal{P}$ because their false numerical noise δ' is generally several orders of magnitude smaller than $\mathcal{P}$, i.e., $|\delta'| \ll |\mathcal{P}|$, and thus is negligible in a long enough time interval $t \in [0, T_c]$, where the critical predictable time T_c is one of the most important concept in CNS. Hence, CNS results are very close to the true physical solutions $\mathcal{P}$ and so can be used as "clean" *benchmark* solutions in $t \in [0, T_c]$. Such benchmark solutions are very useful, as shown in Chapters 4–7 of this book.

Firstly, the numerical noise of a CNS benchmark solution can be much smaller than microscopic physical uncertainty. So, by means of CNS, one can accurately study the propagation and evolution of micro-level physical uncertainty in certain dynamic systems. CNS has highlighted that micro-level physical uncertainty might be the origin of macroscopic randomness of some nonlinear dynamic systems [39, 40, 42, 44]. For further details, please see Chapter 4 of this book.

Secondly, using such "clean" benchmark solutions $\mathcal{S}$ in a long enough time interval $t \in [0, T_c]$, one can accurately investigate the influence of small disturbances on the statistics of chaos through verifying the hypothesis

$$\langle \mathcal{P} + \delta' \rangle = \langle \mathcal{P} \rangle, \quad \text{when } \delta' \sim \mathcal{P} \text{ mostly,} \tag{2.43}$$

where $\langle \rangle$ denotes a statistical operator, and $\mathcal{P} + \delta'$ denotes an another numerical simulation $\mathcal{S}'$ (i.e., $\mathcal{S}'$ is a mixture of $\mathcal{P}$ and δ') given by conventional algorithms using single (or double) precision floating-point arithmetic. The statistics $\langle \mathcal{P} \rangle$ in (2.43) can be accurately estimated from the "clean" benchmark solution $\mathcal{S}$ given by CNS because the CNS trajectory $\mathcal{S}$ is very close to the true physical solution $\mathcal{P}$, say, $\mathcal{S} \approx \mathcal{P}$ is a very good approximation. Thus, CNS provides us with a simple way to verify hypothesis (2.43) for various types of chaotic systems. It has been discovered by means of CNS that the hypothesis is *not* always true: statistics of some chaotic systems are unstable, i.e., sensitive to small disturbances [50, 54, 56, 57]. This has led to a new classification of chaos [54] into *ultra-chaos*, whose statistics are unstable, and *normal-chaos*, whose statistics are stable. Hypothesis (2.43) has a close relationship to reproducibility which is a cornerstone of modern science, and therefore has very important implications. For more details, please see Chapter 5 of this book.

Direct numerical simulation (DNS) [16,17] has been widely used to numerically simulate turbulent flows. By comparing the CNS benchmark solution with the corresponding DNS result, it was found that numerical noises as tiny artificial stochastic disturbances could lead to large-scale deviations of simulations not only in spatiotemporal trajectories but also even in statistics [55]. The essential difference between CNS and DNS is that DNS has no ideas about the critical predictable time T_c, because DNS agrees tacitly that hypothesis (2.43) is always valid, but CNS disagrees it! For further details, please see Chapter 6 of this book.

Unlike other methods, CNS provides us with convergent (and thus reliable) trajectories of chaotic systems. This is especially important for many chaotic systems including the famous three-body problem, where prediction of an accurate orbit over a long time duration is a key objective. By means of CNS, thousands upon thousands of new families of periodic orbits of the three-body system have been found [45–48,52,53]. For more details, please see Chapter 7 of this book. All of the foregoing illustrate that CNS provides us with a new, powerful tool by which to investigate a multitude of open problems.

3

CNS Algorithms for Spatiotemporal Chaos

This chapter briefly describes the basic principles of Clean Numerical Simulation (CNS) [38–40, 43, 54] for spatiotemporal chaos and illustrates some of its applications [49–51, 55–58].

Let us consider a spatiotemporal chaos governed by

$$\dot{u}(x,t) = \mathcal{N}[u(x,t), x, t], \quad x \in [a, b], t \geq 0, \tag{3.1}$$

subject to the initial condition

$$u = f(x), \quad \text{when } t = 0, \tag{3.2}$$

and a periodic boundary condition

$$u\big|_{x=a} = u\big|_{x=b}, \tag{3.3}$$

where x and t are distance and time, $u(x,t)$ is an unknown function dependent on x and t, the dot denotes partial differentiation with respect to t, $\mathcal{N}$ is a nonlinear operator, $[a, b]$ denotes the space domain, and $f(x)$ is a given function.

Our purpose is to obtain a convergent computer-generated simulation $\mathcal{S} = \mathcal{P} + \delta'$ over a prescribed long time interval $t \in [0, T_c]$ during which the false numerical noise δ' is much lower than the true physical solution $\mathcal{P}$, say, $\delta' \ll \mathcal{P}$, so that $\mathcal{S} = \mathcal{P} + \delta'$ in $t \in [0, T_c]$ is an accurate approximation of the true physical solution $\mathcal{P}$, i.e., $\mathcal{S} \approx \mathcal{P}$.

3.1 Basic Principles of CNS for Spatiotemporal Chaos

CNS is based on such the hypothesis that the amplitude of numerical noise (on average) increases exponentially within a given interval of time $t \in [0, T_c]$, i.e.,

$$\mathcal{E}(t) = \mathcal{E}_0 \exp(\kappa t), \quad t \in [0, T_c], \tag{3.4}$$

where the constant $\kappa > 0$ is called the "noise-growth exponent", $\mathcal{E}_0$ denotes the amplitude of background numerical noise, and $\mathcal{E}(t)$ is the amplitude of the

DOI: 10.1201/9781003299622-3

evolving deviation of numerical trajectory from its true physical solution $\mathcal{P}$, and T_c is the "critical predictable time" beyond which the false numerical noise would be at the same order of magnitude as the true physical solution $\mathcal{P}$. In theory, the critical predictable time T_c is determined according to a critical level of noise $\mathcal{E}_c$, which is nearly of the same order of magnitude as the true physical solution $\mathcal{P}$, such that

$$\mathcal{E}_c = \mathcal{E}_0 \exp(\kappa T_c), \tag{3.5}$$

which gives

$$T_c = \frac{1}{\kappa} \ln\left(\frac{\mathcal{E}_c}{\mathcal{E}_0}\right). \tag{3.6}$$

Obviously, for a fixed value of $\mathcal{E}_c$, the lower the amplitude of background numerical noise $\mathcal{E}_0$, the larger the critical predictable time T_c. So, the key strategy of CNS is to reduce background numerical noise $\mathcal{E}_0$ to below a required arbitrarily tiny level.

Note that the background numerical noise $\mathcal{E}_0$ is determined by the maximum of the truncation error[*] and round-off error[†]. So, in the frame of CNS, we must maintain a synchronized decrease in *both* truncation error and round-off error so as to reduce the background numerical noise to below a required arbitrarily small level, as described below.

3.1.1   How to Reduce Background Numerical Noise

The spatial interval $x \in [a, b]$ is discretized into $(N + 1)$ equidistant points, say,

$$x_j = a + \frac{j(b - a)}{N}, \qquad 0 \le j \le N, \tag{3.7}$$

where N is an even number. For a periodic boundary condition (3.3), the unknown continuous function $u(x, t)$ is approximated by a set of N discretized variables

$$\mathbf{U}(t) = \left\{ u(x_0, t), u(x_1, t), u(x_2, t), \cdots, u(x_{N-1}, t) \right\},$$

and the original equation (3.1) is approximated by a finite number of nonlinear ordinary differential equations (ODEs):

$$\dot{u}(x_j, t) = \mathcal{N}[u(x, t), x, t]\Big|_{x=x_j}, \qquad x_j \in [a, b], \ 0 \le j \le N - 1, \tag{3.8}$$

subject to the initial condition

$$u(x_j, 0) = f(x_j), \qquad x_j \in [a, b], \ 0 \le j \le N - 1. \tag{3.9}$$

In this way, we transform the original partial differential equation (3.1) into a finite number of ordinary differential equations, which in principle can be

---

[*]https://en.wikipedia.org/wiki/Truncation_error
[†]https://en.wikipedia.org/wiki/Round-off_error

solved in a similar way to that described in Chapter 2 of this book for temporal chaos.

In the temporal dimension, we use the high-order Taylor expansion

$$u(x_j, t + \Delta t) \approx u(x_j, t) + \sum_{m=1}^{M} u^{[m]}(x_j, t)(\Delta t)^m, \quad 0 \le j \le N - 1, \tag{3.10}$$

where Δt is a time-step and

$$u^{[m]}(x_j, t) = \frac{1}{m!} \left. \frac{\partial^m u(x, t)}{\partial t^m} \right|_{x=x_j}. \tag{3.11}$$

Obviously, for a given time-step Δt, the order M of the temporal Taylor expansion (3.10) must be sufficiently large to reduce the temporal truncation-error to below a required arbitrarily small level. Differentiating $(m - 1)$ times both sides of the governing equation (3.8) with respect to t and then dividing by $m!$, one obtains the following mth-order $(m \ge 2)$ Taylor expansion in the temporal dimension

$$u^{[m]}(x_j, t) = \frac{1}{m!} \left. \frac{\partial^{m-1} N[u(x, t), x, t]}{\partial t^{m-1}} \right|_{x=x_j}, \quad \forall x_j \in [a, b]. \tag{3.12}$$

Unlike temporal chaos, spatial partial derivatives usually occur in (3.12), such as $u_x^{[n]}(x_j, t)$, $u_{xx}^{[n]}(x_j, t)$, etc., where $0 \le n \le m - 1$ and $0 \le j \le N - 1$ are integers. To accurately evaluate these spatial partial derivative terms from the set of *known* discrete variables $u^{[n]}(x_j, t)$, where $0 \le j \le N - 1$, we apply a spectral method [16, 61] in the spatial domain, such as the following Fourier expansion:

$$u^{[n]}(x, t) \approx \sum_{k=-\frac{N}{2}}^{\frac{N}{2}-1} c_k^{[n]}(t) \, e^{ik\alpha x}, \tag{3.13}$$

where $\mathbf{i} = \sqrt{-1}$, $\alpha = 2\pi/(b - a)$ and the coefficients

$$c_k^{[n]}(t) \approx \frac{1}{N} \sum_{j=0}^{N-1} u^{[n]}(x_j, t) \, e^{-ik\alpha x_j}, \quad -\frac{N}{2} \le k \le \frac{N}{2} - 1, \tag{3.14}$$

are determined by known values of $u^{[n]}(x_j, t)$ at the discrete points $x_j \in [a, b]$. Note that N should be a positive even number in this case. Then, the spatial partial derivatives are

$$u_x^{[n]}(x_j, t) \approx \mathbf{i}\,\alpha \sum_{k=-\frac{N}{2}}^{\frac{N}{2}-1} k \, c_k^{[n]}(t) \, e^{ik\alpha x_j}, \quad 0 \le j \le N - 1, \tag{3.15}$$

and

$$u_{xx}^{[n]}(x_j, t) \approx -\alpha^2 \sum_{k=-\frac{N}{2}}^{\frac{N}{2}-1} k^2 \, c_k^{[n]}(t) \, e^{\mathrm{i}k\,\alpha\,x_j}, \qquad 0 \leq j \leq N-1, \qquad (3.16)$$

and so on. Here, the Fast Fourier Transform (FFT) [72] should be used to increase computational efficiency, provided the discrete points x_j are equidistant. In this way, one can reduce the truncation error in the spatial dimension to below a required arbitrarily small level, so long as the number N of discretized points is sufficiently large.

Note that, provided the order M of the Taylor expansion (3.10) is high enough, the magnitude of the *temporal* truncation error is forced below a required arbitrarily small level. Besides, if the spatial discretization is sufficiently fine and the mode number N is large enough, then the *spatial* truncation error of the Fourier spectral expression (3.13) and its spatial derivative terms $u_x(x_k, t)$, $u_{xx}(x_k, t)$, etc., lies below a required arbitrarily tiny level.

In addition, *all* floating-point data are represented using multiple-precision (MP) arithmetic[‡] with a sufficiently large number N_s of significant digits so that the round-off error can be reduced below a required arbitrarily tiny level. In this way, *both* round-off and spatiotemporal truncation errors can be reduced to below a required arbitrarily small level.

To produce convergent chaotic simulations efficiently, the "variable step-size variable order" (VSVO) scheme [67] is applied in the temporal dimension for a given allowable tolerance *tol* of the governing equation (3.8). By combining the high-order Taylor expansion method with the VSVO scheme, high precision calculations can be optimized. According to Barrio *et al.* [67], the optimal time-step is given by

$$\Delta t = \min\left(\frac{tol^{\frac{1}{M}}}{\left\| u^{[M-1]}(x_j, t) \right\|_\infty^{\frac{1}{M-1}}}, \frac{tol^{\frac{1}{M+1}}}{\left\| u^{[M]}(x_j, t) \right\|_\infty^{\frac{1}{M}}} \right), \quad \forall x_j \in [a, b], \qquad (3.17)$$

where $\| \ \|_\infty$ is the infinity norm, *tol* is the permitted tolerance of the governing equation (3.8), M is the order of the Taylor expansion (3.10) in the temporal dimension, and $x_j \in [a, b]$ with $0 \leq j \leq N-1$. Besides, in order to further increase computational efficiency, the round-off error should be of the same level as the (temporal) truncation error, so that

$$tol = 10^{-N_s} \qquad (3.18)$$

should hold, where N_s is the number of significant digits used in the multiple precision (MP) representation of all floating-point data. Substituting (3.18)

[‡]The multiple-precision [62] representation is freely available and easy to install. For details, please visit `https://gmplib.org/` to install the GMP (GNU multiple precision arithmetic library) and `https://mpfr.loria.fr/` to install the GNU MPRF library.

into (3.17), the optimal time-step is given by

$$\Delta t = \min\left(\frac{10^{-\frac{N_s}{M}}}{\left\|u^{[M-1]}(x_j,t)\right\|_\infty^{\frac{1}{M-1}}}, \frac{10^{-\frac{N_s}{M+1}}}{\left\|u^{[M]}(x_j,t)\right\|_\infty^{\frac{1}{M}}}\right), \quad \forall x_j \in [a,b]. \qquad (3.19)$$

Note that, in the VSVO scheme [67], one has the freedom to choose a proper value of M for a given *tol*. Using (3.18), we could adopt the following empirical formula

$$M = -c_1^E \log_{10}(tol) + c_0^E = c_1^E N_s + c_0^E, \qquad (3.20)$$

where c_0^E and c_1^E are empirical constants.

In this way, both the round-off error and spatiotemporal truncation error can be reduced to below a required arbitrarily tiny level by choosing both a sufficiently large mode number N for the spatial Fourier expansion (3.13) and a sufficiently large number N_s of significant digits for the multiple precision arithmetic.

3.1.2   Determination of Critical Predictable Time T_c

Let $u(x,t)$ denote a numerical simulation given by CNS using the spatial Fourier expansion (3.13) with mode number N, and N_s be the number of significant digits used in the multiple precision arithmetic. Now, let $u'(x,t)$ be another numerical simulation by CNS with even smaller background numerical noise, determined by a new mode number N' for the spatial Fourier expansion and a new number N_s' of significant digits of the multiple precision, where $N' > N$ and $N_s' \geq N_s$. Since the numerical noise increases exponentially, the simulation $u'(x,t)$ with lower background numerical noise should be closer to the true physical solution $\mathcal{P}$ over a longer interval of time, and thus can be used as a benchmark solution by which to calculate the critical predictable time T_c.

Given that $u'(x,t)$ might have a different spatial mode number from $u(x,t)$, it is convenient to compare their spatial spectra. Rewriting $u(x,t)$ and $u'(x,t)$ in terms of spatial Fourier expressions, we have

$$u(x,t) \approx \sum_{k=-\frac{N}{2}}^{\frac{N}{2}-1} c_k(t)\, e^{\mathbf{i} k \alpha x}, \quad u'(x,t) \approx \sum_{k=-\frac{N'}{2}}^{\frac{N'}{2}-1} c_k'(t)\, e^{\mathbf{i} k \alpha x}, \qquad (3.21)$$

where $\mathbf{i} = \sqrt{-1}$ and $\alpha = 2\pi/(b-a)$. Generally, the spectrum energy $\sum\limits_{k=-\frac{N'}{2}}^{\frac{N'}{2}-1} \left|c_k'\right|^2$

has physical meaning. So, from a physical standpoint, it is important for any numerical simulation of spatiotemporal chaos to have an accurate spatial spectrum. With this in mind, we define the "spectrum-deviation" as

$$\delta_s(t) = \frac{\sum\limits_{k=-\frac{N}{2}}^{\frac{N}{2}-1} \left| |c'_k|^2 - |c_k|^2 \right|}{\sum\limits_{k=-\frac{N'}{2}}^{\frac{N'}{2}-1} |c'_k|^2} \tag{3.22}$$

in order to quantify the difference between $u(x,t)$ and $u'(x,t)$ at a given time t. Obviously, the smaller the spectrum-deviation δ_s, the better the two numerical simulations agree with each other. So, it is reasonable to define a *convergence criterion* by

$$\delta_s \le \delta_s^c, \tag{3.23}$$

where $\delta_s^c > 0$ is a small number, called the "critical spectrum-deviation". From the author's experience, $\delta_s^c = 0.01$ is reasonable for many problems [49, 50]. Note that (3.23) provides a convergence criterion, which determines the "critical predictable time" T_c and corresponding temporal interval $[0, T_c]$ during which the convergence criterion (3.23) is satisfied and the difference between $u(x,t)$ and $u'(x,t)$ is negligible.

Our experience [49–51] suggests that

$$T_c \approx a_1^R N_s + a_0^R, \tag{3.24}$$

and

$$T_c \approx \mu_1^R N + \mu_0^R, \tag{3.25}$$

where N is the mode number of the spatial spectral expansion, N_s is the number of significant digits of the multiple precision representation of all floating-point data, and $a_1^R, a_0^R, \mu_1^R,$ and μ_0^R are regression coefficients determined by a few test simulations using different values of N and N_s. Thus, for spatiotemporal chaos, the critical predictable time T_c of a numerical simulation linearly increases with respect to the mode number N of the spatial spectral expansion and the number N_s of significant digits of multiple precision arithmetic used for all floating-point data. In practice, we have

$$T_c \approx \frac{1}{\gamma} \min\left\{ a_1^R N_s + a_0^R, \ \mu_1^R N + \mu_0^R \right\}, \tag{3.26}$$

where $\gamma \ge 1$ is a constant, which is used here as a kind of safety factor.

Thus, given a critical predictable time T_c, we have the corresponding number of significant digits required by the multiple precision arithmetic as

$$N_s = \left\lceil \frac{\gamma \, T_c - a_0^R}{a_1^R} \right\rceil \tag{3.27}$$

and the required mode number of the spatial spectral expansion as

$$N = \left\lceil \frac{\gamma \, T_c - \mu_0^R}{\mu_1^R} \right\rceil, \tag{3.28}$$

where $\lceil x \rceil$ is an operator that takes the integer part of x, and $\gamma \geq 1$ is used here as a safety factor.

3.1.3   Self-adaptive CNS Algorithm with Variable Precision

According to (3.4), the numerical noise $\mathcal{E}(t)$ increases *exponentially*. Thus, after a time duration t^*, $\mathcal{E}(t^*)$ becomes much larger than the background numerical noise $\mathcal{E}_0$. So, it is *unnecessary* to keep the background numerical noise $\mathcal{E}_0$ fixed throughout the whole interval $t \in [0, T_c]$. In theory, according to (3.4), use of a larger background numerical noise $\mathcal{E}_0^* > \mathcal{E}_0$ does not influence the numerical result in $t > t^*$, so long as $\mathcal{E}_0^* \leq \mathcal{E}(t^*)$.

So, in order to increase computational efficiency, even if the mode number N of the spatial spectral expansion is fixed, one could gradually decrease the number N_s of significant digits of the multiple precision arithmetic as the integration time t increases. For CNS algorithms with an optimal variable time-step, according to (3.27), it is sufficient, after integrating over a time duration $t^* < T_c$, to use a smaller number N_s of significant digits of the multiple precision arithmetic, such that

$$N_s = \left\lceil \frac{\gamma \, (T_c - t^*) - a_0^R}{a_1^R} \right\rceil, \qquad \text{when } t^* \leq t \leq T_c, \tag{3.29}$$

where $\lceil x \rceil$ is an operator that takes the integer part of x, $\gamma \geq 1$ is a constant acting as a kind of safety factor, and a_0^R and a_1^R are regression coefficients.

The aforementioned self-adaptive CNS algorithm with gradually decreasing multiple-precision and variable optimal time-step can greatly increase the computational efficiency of CNS algorithms for spatiotemporal chaos, as illustrated below. For details, please refer to Qin and Liao [58].

3.2   An Illustrative Example: Sine-Gordon Equation

Without loss of generality, let us consider the continuum limit of a chain of pendulums coupled through an elastic restoring force and damped friction, governed by the damped driven sine-Gordon equation [73–75]:

$$u_{tt} = u_{xx} - \sin(u) - \alpha u_t + \Gamma \sin(\omega t - \lambda x), \tag{3.30}$$

subject to the periodic boundary condition

$$u(x + l, t) = u(x, t), \tag{3.31}$$

where u denotes the angle corresponding to the vertical direction of the pendulum, t and x are temporal and spatial variables, the subscript denotes the partial derivative, α and Γ are physical parameters, ω is the temporal frequency, and $\lambda = 2\pi/l$ with l being the total length of the system. Using the transformation $x' = \lambda x + \pi$, the governing equation (3.30) becomes

$$u_{tt} = \lambda^2 u_{xx} - \sin(u) - \alpha u_t - \Gamma \sin(\omega t - x), \tag{3.32}$$

where $x \in [0, 2\pi]$ is the spatial variable and the prime is neglected for simplicity, subject to the periodic boundary condition

$$u(0, t) = u(2\pi, t). \tag{3.33}$$

Without loss of generality, let us consider the initial condition

$$u(x, 0) = 0, \qquad u_t(x, 0) = 0, \tag{3.34}$$

for a case with the following parameters

$$\omega = \frac{3}{5}, \qquad \alpha = \frac{1}{10}, \qquad \Gamma = \frac{461}{500}, \qquad l = 500, \tag{3.35}$$

corresponding to a spatiotemporal chaos [74] solved by Qin and Liao [50] using CNS.

3.2.1 CNS Algorithm for Spatiotemporal Chaos

First, the spatial interval $x \in [0, 2\pi]$ is discretized into $N+1$ equidistant points

$$x_k = \frac{2\pi k}{N}, \qquad k = 0, 1, 2, ..., N. \tag{3.36}$$

Then, $u(x, t)$ is approximated by a set of N unknown time-dependent functions

$$\mathbf{U}(t) = \left\{ u(x_0, t), u(x_1, t), u(x_2, t), \cdots, u(x_{N-1}, t) \right\}.$$

In the temporal dimension, the Mth-order Taylor expansion

$$u(x_k, t + \Delta t) \approx u(x_k, t) + \sum_{m=1}^{M} u^{[m]}(x_k, t)(\Delta t)^m \tag{3.37}$$

is used for each $x_k \in [0, 2\pi]$, where Δt denotes the time-step and

$$u^{[m]}(x_k, t) = \frac{1}{m!} \frac{\partial^m u(x_k, t)}{\partial t^m} \tag{3.38}$$

is the mth-order series of temporal partial derivatives. For a given time-step Δt, the order M must be sufficiently large as to reduce the truncation error (in the temporal dimension) to below a required arbitrarily tiny level.

Differentiating $(m-2)$ times $(m > 2)$ both sides of Eq. (3.32) with respect to t and then dividing by $m!$, one obtains the high-order temporal derivatives as

$$u^{[m]}(x_k,t) = \frac{1}{(m-1)m}\left[\lambda^2 u_{xx}^{[m-2]}(x_k,t) - W_{m-2}(x_k,t)\right]$$
$$-\left(\frac{\alpha}{m}\right)u^{[m-1]}(x_k,t) - \left(\frac{\Gamma \omega^{m-2}}{m!}\right)F_{m-2}(x_k,t), \qquad (3.39)$$

where

$$F_n(x_k,t) = \begin{cases} +\sin(\omega t - x_k), & n = 4r, \\ +\cos(\omega t - x_k), & n = 4r+1, \\ -\sin(\omega t - x_k), & n = 4r+2, \\ -\cos(\omega t - x_k), & n = 4r+3, \end{cases} \qquad (3.40)$$

for integers $n, r \geq 0$, and $W_{m-2}(x_k,t)$ is determined by the following recursion formulas

$$\begin{cases} W_0(x_k,t) = \sin[u(x_k,t)], \\ V_0(x_k,t) = \cos[u(x_k,t)], \\ W_j(x_k,t) = \frac{1}{j}\sum_{i=0}^{j-1}(j-i)u^{[j-i]}(x_k,t)\,V_i(x_k,t), & j \geq 1, \\ V_j(x_k,t) = -\frac{1}{j}\sum_{i=0}^{j-1}(j-i)u^{[j-i]}(x_k,t)\,W_i(x_k,t), & j \geq 1. \end{cases} \qquad (3.41)$$

Besides, for a spatiotemporal chaos, there always exist terms with spatial derivatives, such as $u_{xx}^{[m-2]}(x_k,t)$ in Eq. (3.39), which must also be evaluated to a high level of accuracy. Note that one of the most accurate approximations of $u^{[j]}(x,t)$ is its Fourier spectral expression

$$u^{[j]}(x,t) \approx \frac{1}{2}a_{j,0}(t) + \sum_{n=1}^{\frac{N}{2}-1}\left[a_{j,n}(t)\,\cos(nx) + b_{j,n}(t)\,\sin(nx)\right]$$
$$+a_{j,\frac{N}{2}}(t)\cos\left(\frac{Nx}{2}\right), \qquad (3.42)$$

where

$$a_{j,n}(t) = \frac{2}{N}\sum_{k=0}^{N-1}u^{[j]}(x_k,t)\cos(n\,x_k), \qquad (3.43)$$

$$b_{j,n}(t) = \frac{2}{N}\sum_{k=0}^{N-1}u^{[j]}(x_k,t)\sin(n\,x_k). \qquad (3.44)$$

Here, the fast Fourier transform (FFT) algorithm is applied. Then, from (3.42), the second derivative becomes

$$u_{xx}^{[j]}(x_k,t) \approx - \sum_{n=1}^{\frac{N}{2}-1} n^2 \left[a_{j,n}(t) \cos(nx_k) + b_{j,n}(t) \sin(nx_k) \right]$$

$$-a_{j,\frac{N}{2}}(t) \left(\frac{N}{2} \right)^2 \cos \left(\frac{N x_k}{2} \right), \tag{3.45}$$

where $j \geq 0$ is an integer and $x_k \in [0, 2\pi]$.

Note that, if the order M of the temporal Taylor expansion (3.37) is high enough, the *temporal* truncation error can be controlled so that it remains below a required tiny level. Besides, if the spatial discretization is sufficiently fine such that the mode number N is sufficiently large, then the *spatial* truncation error of the Fourier spectral expression (3.42) and the corresponding spatial derivative term $u_{xx}^{[m-2]}(x_k,t)$ given by (3.45) can be restricted to lie below a required arbitrarily tiny level. Furthermore, by representing all physical/numerical variables and parameters using multiple precision (MP) with a sufficiently large number N_s of significant digits, the round-off error can be reduced to below a required arbitrarily tiny level. In this way, *both* the round-off error and spatiotemporal truncation error can be reduced below a required arbitrarily tiny level.

In order to gain reliable simulations efficiently for a given allowable tolerance *tol*, the variable stepsize variable order (VSVO) scheme [67] is applied in the temporal dimension to solve the set of N ordinary differential equations for the time-dependent unknown function $u(x_k,t)$, where $k = 0, 1, 2, \cdots, N-1$, and $x_k \in [0, 2\pi]$ is the coordinate at each discretized point x_k, by means of the following optimal time-step

$$\Delta t = \min \left(\frac{tol^{\frac{1}{M}}}{\left\| u^{[M-1]}(x_k,t) \right\|_{\infty}^{\frac{1}{M-1}}}, \frac{tol^{\frac{1}{M+1}}}{\left\| u^{[M]}(x_k,t) \right\|_{\infty}^{\frac{1}{M}}} \right), \quad \forall x_k \in [0, 2\pi], \tag{3.46}$$

where M is the order of Taylor expansion (3.37) in the temporal dimension, $\| \ \|_{\infty}$ is the infinity norm, and *tol* is the allowable tolerance. Note that, for a given allowable tolerance *tol*, one has freedom to choose a proper value for the order M of the Taylor expansion to achieve high computational efficiency. Here, let us adopt the following empirical formula

$$M = -\log(tol) - 10. \tag{3.47}$$

In order to increase computational efficiency, the round-off error should be of the same level as the temporal truncation error, so that

$$tol = 10^{-N_s}, \tag{3.48}$$

where N_s is the number of significant digits in multiple precision arithmetic for all floating-point data. Substituting (3.48) into (3.47) gives the empirical formula

$$M = N_s - 10 \tag{3.49}$$

for our problem of interest. For more details, please refer to Qin and Liao [50].

3.2.2 Determination of Critical Predictable Time T_c

Let $u(x,t)$ (corresponding to a mode-number N of the spatial Fourier expansion) and $u'(x,t)$ (corresponding to another mode-number N' of the spatial Fourier expansion) denote two CNS results with different levels of background numerical noise, where $N' > N$, say, $u'(x,t)$ has lower numerical noise than $u(x,t)$, provided both are gained using the multiple precision arithmetic with the same number of significant digits. Given that numerical noise increases exponentially, causing the chaotic trajectory to depart from its true physical solution $\mathcal{P}$ exponentially, then $u'(x,t)$ should be closer to $\mathcal{P}$ and thus should depart from $\mathcal{P}$ later than $u(x,t)$. So, $u'(x,t)$ can be used as a benchmark solution to determine the critical predictable time T_c, before which (i.e., $t \le T_c$) $u(x,t)$ and $u'(x,t)$ agree well with each other and thus can be regarded as "convergent". At any given time t, both $u(x,t)$ and $u'(x,t)$ can be expressed by the spatial Fourier expansion (3.42), with a_n, b_n, a'_n and b'_n denoting the corresponding coefficients. Writing

$$c_n^2 = a_{0,n}^2 + b_{0,n}^2, \qquad (c'_n)^2 = (a'_{0,n})^2 + (b'_{0,n})^2, \tag{3.50}$$

then, the spectrum-deviation, given by

$$\delta_s(t) = \frac{\sum\limits_{n=0}^{\frac{N}{2}} \left| (c'_n)^2 - c_n^2 \right|}{\sum\limits_{n=0}^{\frac{N'}{2}} (c'_n)^2}, \tag{3.51}$$

provides a measure of the deviation of $u(x,t)$ from the benchmark solution $u'(x,t)$. Finally, the critical predictable time T_c of $u(x,t)$ (carrying the higher background numerical noise) is determined by the convergence criterion

$$\delta_s(t) \le \delta_s^c, \tag{3.52}$$

where $\delta_s^c > 0$ is a small constant. According to the author's experience, $\delta_s^c = 1\%$ is generally sufficient in practice, and is used here to determine T_c.

Using the temporal allowable tolerance $tol = 10^{-60}$ and setting the number of significant digits to $N_s = 60$ for the multiple-precision arithmetic, it is found that the critical predictable time T_c increases almost linearly with respect to the mode number N of the spatial Fourier expansion, such that

$$T_c \approx 0.0558N + 10.7. \tag{3.53}$$

Besides, if N is sufficiently large (e.g., $N = 16384$) and *tol* is given by (3.48), then the critical predictable time T_c is found to increase almost linearly with respect to N_s, such that

$$T_c \approx 16.5 N_s - 87.8. \tag{3.54}$$

Therefore, we have

$$T_c \approx \frac{1}{\gamma} \min \left\{ 0.0558 N + 10.7, 16.5 N_s - 87.8 \right\}, \tag{3.55}$$

where $\gamma \geq 1$ is a kind of safety factor.

Then, according to (3.55), one can gain a reliable/convergent simulation of spatiotemporal chaos over a prescribed long time interval $t \in [0, 900]$ by setting the mode number of the spatial Fourier expansion to $N = 16384$, the number of significant digits in multiple precision floating-point arithmetic to $N_s = 60$, and the temporal allowable tolerance to *tol* $= 10^{-60}$ for the high-order Taylor expansion, if we set $\gamma = 1$. To verify its convergence and reliability, we compare this CNS result with another CNS result for even *lower* background numerical noise, obtained for the same mode number $N = 16384$ but *smaller* temporal allowable tolerance *tol* $= 10^{-70}$ and *higher* multiple precision floating-point arithmetic with more significant digits (say $N_s = 70$). It is then found that the corresponding spectrum-deviation δ_s defined by (3.51) between the two CNS results is indeed less than 1% over the time interval $0 \leq t \leq 900$. This verifies the convergence and reliability of the CNS result for $N = 16384$, *tol* $= 10^{-60}$ and $N_s = 60$. For more details, please refer to Qin and Liao [50].

3.2.3   Self-adaptive CNS Algorithm with Variable Precision

According to (3.53) and (3.54), for a specified critical predictable time T_c concerning the spatiotemporal chaos under consideration, the required mode number of the spatial Fourier expansion is

$$N \approx \left\lceil 17.92 \, \gamma \, T_c - 191.8 \right\rceil, \tag{3.56}$$

and the required number of significant digits in floating-point multiple precision is

$$N_s \approx \left\lceil 0.061 \, \gamma \, T_c + 5.32 \right\rceil, \tag{3.57}$$

where $\lceil x \rceil$ is an operator that takes the integer part of x, and $\gamma \geq 1$ is the safety factor.

Note that the background numerical noise $\mathcal{E}_0$ over the whole interval of time $t \in [0, T_c]$ is determined by N and N_s as given by the above expressions. However, it is *unnecessary* to fix the number N_s of significant digits in multiple precision arithmetic throughout the *whole* interval $t \in [0, T_c]$: according to (3.4), the numerical noise $\mathcal{E}(t)$ increases exponentially so that, after a time duration t^*, the numerical noise $\mathcal{E}(t)$ becomes much larger than the background numerical noise $\mathcal{E}_0$ when $t^* \leq t \leq T_c$. Thus, a controlled increase in background numerical noise should not influence the numerical simulation, provided the background numerical noise remains below $\mathcal{E}(t^*)$. Therefore, after a time duration t^*, one could use the same mode number N in the spatial Fourier expansion but a smaller number N_s of significant digits in the multiple precision floating-point arithmetic, such that

$$N_s \approx \left\lceil 0.061 \, \gamma \, (T_c - t^*) + 5.32 \right\rceil, \tag{3.58}$$

where $t \in [t^*, T_c]$.

In practice, one can fix the mode number N in the spatial Fourier expansion but decrease the number of significant digits N_s of all floating-point data at $t^* = n\Delta T$, where $n \geq 1$ is an integer and ΔT is a constant (such as $\Delta T = 25$, 50, 100, etc.). Use of this kind of self-adaptive CNS algorithm can greatly increase the computational efficiency. For more details, please refer to Qin and Liao [58].

4

On the Origin of Macroscopic Randomness

It is widely accepted that the microscopic world is random. Moreover, randomness and uncertainty are ubiquitous in the macroscopic world. Two questions immediately arise. Does macroscopic randomness have any relationship with micro-level uncertainty? What are the origins of macroscopic randomness and uncertainty?

Due to the butterfly effect, any tiny disturbance of a chaotic system grows exponentially. Thus, a conventional numerical simulation of chaos comprises a mixture of the true physical solution $\mathcal{P}$ and false numerical noise δ', which are mostly of the same order of magnitude, say $\delta' \sim \mathcal{P}$. As a result, the false numerical noise δ' might be mostly much larger than micro-level physical uncertainty ϵ, i.e., $|\delta'| \gg |\epsilon|$. In such cases, it is impossible to investigate accurately the influence of micro-level physical uncertainty on the chaotic trajectory and its statistics.

Unlike conventional schemes, Clean Numerical Simulation (CNS) [38–40, 43,54] can provide us with a benchmark trajectory S over a prescribed sufficiently long time interval $t \in [0, T_c]$, during which the false numerical noise δ' is always much smaller than micro-level physical uncertainty ϵ and the true physical solution $\mathcal{P}$. In short, $|\delta'| \ll |\epsilon| \ll \mathcal{P}$. And so $S = \mathcal{P} + \delta'$ is very close to the true physical solution $\mathcal{P}$, i.e., $S \approx \mathcal{P}$. Thus, one can accurately investigate the propagation of micro-level physical uncertainty ϵ and more importantly its influence on the statistical results.

In this chapter, using CNS [38–58] as a powerful tool, Newton's three-body problem and the Rayleigh-Bénard convection turbulence are used to provide theoretical evidence that the origin of macroscopic randomness and uncertainty might lie in micro-level physical uncertainty.

4.1 A Chaotic Three-Body System

4.1.1 Mathematical Model and the CNS Algorithm

Consider Newton's three-body problem expressed as the motion of three celestial bodies under mutual gravitational attraction. Let x_1, x_2, x_3 denote three orthogonal axes. The position vector of the ith body is expressed by

DOI: 10.1201/9781003299622-4

$\mathbf{r}_i = (x_{1,i}, x_{2,i}, x_{3,i})$. Let T and L denote characteristic time and length scales, and m_i the mass of the ith body. Using Newton's law of gravitational attraction and Newton's second law of motion, the movement of the three bodies is governed by the following non-dimensional equations

$$\ddot{x}_{k,i} = \sum_{j=1, j \neq i}^{3} \rho_j \frac{(x_{k,j} - x_{k,i})}{R_{i,j}^3}, \qquad k, i = 1, 2, 3, \qquad (4.1)$$

subject to a given initial condition

$$\mathbf{r}_i(0) = \mathbf{r}_{i,0}, \quad \dot{\mathbf{r}}_i(0) = \dot{\mathbf{r}}_{i,0}, \qquad (4.2)$$

in which

$$\rho_i = \frac{m_i}{m_1}, \quad i = 1, 2, 3, \qquad (4.3)$$

is the mass ratio, $\mathbf{r}_{i,0}$ is the given initial position, $\dot{\mathbf{r}}_{i,0}$ is the given initial velocity of the ith body, and

$$R_{i,j} = \left[\sum_{k=1}^{3} (x_{k,j} - x_{k,i})^2 \right]^{1/2}. \qquad (4.4)$$

In the frame of CNS, the Mth-order Taylor expansion,

$$x_{k,i}(t + \Delta t) \approx \sum_{m=0}^{M} \alpha_m^{k,i} (\Delta t)^m, \qquad (4.5)$$

is used to calculate accurately the orbits of the three bodies, where $\alpha_m^{k,i}$ is a constant coefficient dependent upon the time t. Note that the position $x_{k,i}(t)$ and velocity $\dot{x}_{k,i}(t)$ at time t are known, i.e.,

$$\alpha_0^{k,i} = x_{k,i}(t), \quad \alpha_1^{k,i} = \dot{x}_{k,i}(t). \qquad (4.6)$$

The recursion formula for $\alpha_m^{k,i}$ when $m \geq 2$ is derived from (4.1), as described below.

Write

$$f_{i,j} = \frac{1}{R_{i,j}^3} \approx \sum_{m=0}^{M} \beta_m^{i,j} (\Delta t)^m, \qquad (4.7)$$

where Δt is a time-step and $\beta_m^{i,j} = \beta_m^{j,i}$. Substituting (4.5) and (4.7) into (4.1) and comparing like-powers of the time-step Δt, we obtain the recursion formula

$$\alpha_{m+2}^{k,i} = \frac{1}{(m+1)(m+2)} \sum_{j=1, j \neq i}^{3} \rho_j \sum_{n=0}^{m} \left(\alpha_n^{k,j} - \alpha_n^{k,i} \right) \beta_{m-n}^{i,j}, \quad m \geq 0. \qquad (4.8)$$

Thus, the positions and velocities of the three bodies at the time $t + \Delta t$ are given by

$$x_{k,i}(t + \Delta t) \approx \sum_{m=0}^{M} \alpha_m^{k,i} (\Delta t)^m, \tag{4.9}$$

$$\dot{x}_{k,i}(t + \Delta t) \approx \sum_{m=1}^{M} m \, \alpha_m^{k,i} (\Delta t)^{m-1}. \tag{4.10}$$

We now write

$$S_{i,j} = R_{i,j}^6 \approx \sum_{m=0}^{M} \gamma_m^{i,j} (\Delta t)^m, \quad f_{i,j}^2 = \sum_{m=0}^{M} \sigma_m^{i,j} (\Delta t)^m, \tag{4.11}$$

which has the symmetry property $\gamma_m^{i,j} = \gamma_m^{j,i}$ and $\sigma_m^{i,j} = \sigma_m^{j,i}$. Substituting (4.4), (4.5) and (4.7) into the above definitions and comparing like-powers of the time-step Δt, we have

$$\gamma_m^{i,j} = \sum_{n=0}^{m} \mu_{m-n}^{i,j} \sum_{k=0}^{n} \mu_k^{i,j} \mu_{n-k}^{i,j}, \tag{4.12}$$

$$\sigma_m^{i,j} = \sum_{n=0}^{m} \beta_n^{i,j} \beta_{m-n}^{i,j}, \tag{4.13}$$

where

$$\mu_m^{i,j} = \sum_{k=1}^{3} \sum_{n=0}^{m} \left(\alpha_n^{k,j} - \alpha_n^{k,i} \right) \left(\alpha_{m-n}^{k,j} - \alpha_{m-n}^{k,i} \right), \quad i \neq j, \; m \geq 1, \tag{4.14}$$

with symmetry $\mu_m^{i,j} = \mu_m^{j,i}$. According to (4.7), this holds for

$$S_{i,j} f_{i,j}^2 = 1.$$

Substituting (4.11) into the above equation and comparing like-powers of Δt, we have

$$\sum_{n=0}^{m} \gamma_n^{i,j} \sigma_{m-n}^{i,j} = 0, \quad m \geq 1,$$

which gives the recursion formula

$$\beta_m^{i,j} = -\frac{1}{2\beta_0^{i,j} \gamma_0^{i,j}} \left\{ \sum_{n=1}^{m} \gamma_n^{i,j} \sigma_{m-n}^{i,j} + \gamma_0^{i,j} \sum_{k=1}^{m-1} \beta_k^{i,j} \beta_{m-k}^{i,j} \right\}, \quad m \geq 1, \tag{4.15}$$

where $i \neq j$. In addition, we have at the time t that

$$\beta_0^{i,j} = \frac{1}{R_{i,j}^3}, \quad \mu_0^{i,j} = R_{i,j}^2, \quad \sigma_0^{i,j} = \left(\beta_0^{i,j} \right)^2, \quad \gamma_0^{i,j} = \left(\mu_0^{i,j} \right)^3. \tag{4.16}$$

Further details are given by Liao [40].

According to (2.22), one can use the optimal variable time-step

$$\Delta t = \min \left(\frac{10^{-\frac{N_s}{M}}}{\left\| \alpha_{M-1}^{k,i} \right\|_{\infty}^{\frac{1}{M-1}}}, \frac{10^{-\frac{N_s}{M+1}}}{\left\| \alpha_{M}^{k,i} \right\|_{\infty}^{\frac{1}{M}}} \right), \qquad 1 \le i, k \le 3, \qquad (4.17)$$

where M is the order of the Taylor expansion (4.5), $\alpha_m^{k,i}$ is the coefficient of the Taylor expansion (4.5), $\| \ \|_{\infty}$ is the infinite norm, and N_s is the number of significant digits in floating-point multiple-precision for all data. The value for the order M is selected from either the optimal expression (2.25) or the empirical formula (2.23). According to (2.28), the number N_s of significant digits in the multiple precision floating-point arithmetic is specified by

$$N_s = \left\lceil \frac{\gamma \, \kappa \, T_c}{\ln 10} - \log_{10} \mathcal{E}_c \right\rceil, \qquad (4.18)$$

where $\mathcal{E}_c$ denotes the critical numerical noise that is nearly of the order of magnitude of the true physical solution $\mathcal{P}$, κ is the noise-growth exponent that is equal to the positive leading Lyapunov exponent, $\lceil x \rceil$ is an operator that takes the integer value of x, and $\gamma \ge 1$ is a constant used as a kind of safety factor.

From (2.32), after a simulation time duration t^*, the number of significant digits can be reduced to

$$N_s = \left\lceil \frac{\gamma \, \kappa (T_c - t^*)}{\ln 10} - \log_{10} \mathcal{E}_c \right\rceil, \qquad \text{for } t^* \le t \le T_c, \qquad (4.19)$$

in order to increase the computational efficiency. For further details, please see Section 2.2.4 of this book about the self-adaptive CNS algorithm with variable precision, and refer to Qin and Liao [58].

4.1.2 From Micro-level Uncertainty to Macroscopic Randomness

Without loss of generality, now consider the motion of three bodies with the three equal masses, i.e., $\rho_1 = \rho_2 = \rho_3 = 1$, subject to the initial positions

$$\mathbf{r}_1 = (0, 0, -1) + d\mathbf{r}_1, \ \mathbf{r}_2 = (0, 0, 0), \mathbf{r}_3 = -(\mathbf{r}_1 + \mathbf{r}_2), \qquad (4.20)$$

and the initial velocities

$$\dot{\mathbf{r}}_1 = (0, -1, 0), \dot{\mathbf{r}}_2 = (1, 1, 0), \dot{\mathbf{r}}_3 = -(\dot{\mathbf{r}}_1 + \dot{\mathbf{r}}_2), \qquad (4.21)$$

where $d\mathbf{r}_1$ denotes a kind of micro-level physical uncertainty, given by

$$d\mathbf{r}_1 = +10^{-60}(1, 0, 0), \ \ d\mathbf{r}_1 = -10^{-60}(1, 0, 0), \ \ d\mathbf{r}_1 = 10^{-60}(1, 1, 1),$$

etc. It should be emphasized that this kind of tiny physical uncertainty is much smaller than numerical noise produced by conventional algorithms based on single (or double) precision floating-point arithmetic. Fortunately, by means of CNS, the numerical noise can be constrained to be much smaller than the tiny level of physical uncertainty over a sufficiently long enough time interval $t \in [0, T_c]$. Note that the initial conditions satisfy

$$\sum_{j=1}^{3} \dot{\mathbf{r}}_j(0) = \sum_{j=1}^{3} \mathbf{r}_j(0) = 0. \tag{4.22}$$

Thus, due to momentum conversation, it invariably holds that

$$\sum_{j=1}^{3} \dot{\mathbf{r}}_j(t) = \sum_{j=1}^{3} \mathbf{r}_j(t) = 0, \qquad t \geq 0. \tag{4.23}$$

As pointed out by Sprott [8], when no micro-level physical uncertainty is present, i.e., $d\mathbf{r}_1 = (0, 0, 0)$, the three-body problem corresponds to a chaotic system with maximum Lyapunov exponent $\lambda = 0.1681$, as shown in Figure 4.1. Note that Body-2 moves along a straight line, which is the axis of symmetry of the chaotic orbits of Body-1 and Body-3.

Convergent, reliable orbits of the aforementioned chaotic three-body system over a prescribed, sufficiently long time interval $t \in [0, T_c]$ with critical predictable time $T_c \geq 1000$ have been obtained by Liao [40] using CNS with 300 significant digits of floating-point multiple-precision arithmetic, a fixed time-step of $\Delta t = 0.01$ (or 0.001), and a Taylor expansion of optimal order M. A variable optimal time-step (4.17) was used to enhance the computational efficiency, where the optimal order,

$$M = \left\lceil 1.15 \, N_s + 1 \right\rceil,$$

is given by (2.25) and the self-adaptive number of significant digits,

$$N_s = \left\lceil 0.08(T_c - t^*) + 2 \right\rceil, \qquad \text{when } t^* \leq t \leq T_c,$$

is given by (4.19) with $T_c - t^* \geq 200$, for $\kappa = 0.1681$, the critical noise $\mathcal{E}_c = 10^{-2}$, and the safety factor $\gamma = 1.1$. All the CNS results reported by Liao [40] and Liao & Li [42] can be reproduced in this way leading to much lower CPU overhead [58]. Liao [40] also found that, when micro-level physical uncertainty exists such that $|d\mathbf{r}_1|$ has an order of magnitude of 10^{-60}, then the corresponding chaotic three-body system has the same maximum Lyapunov exponent, i.e., $\lambda = 0.1681$. However, when $|d\mathbf{r}_1| \neq 0$, the trajectory symmetries of Body-1 and Body-3 with respect to Body-2 break down. For example, when $d\mathbf{r}_1 = 10^{-60}(1, 1, 1)$, the orbits of the corresponding three-body system appear to be the same as those without micro-level uncertainty (i.e., $|d\mathbf{r}_1| = 0$)

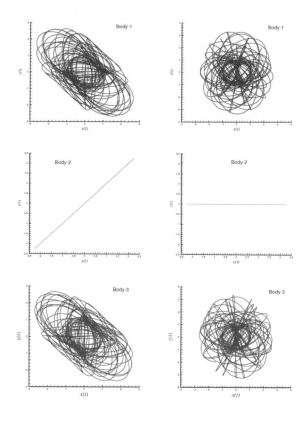

FIGURE 4.1

Convergent trajectories of the three-body system with three equal masses during $t \in [0, 1000]$ in the $x - y$ plane (left) and the $x - z$ plane (right) for initial conditions given by (4.20) and (4.21) when $d\mathbf{r}_1 = (0, 0, 0)$. Upper plot: Body-1; middle plot: Body-2; and lower plot: Body-3.

during the time interval $t \in [0, T^*]$ where $T^* \approx 800$. However, the deviation becomes increasingly obvious for $t > T^*$, as shown in Figure 4.2. Here, we call T^* the "propagation time of microscopic uncertainty to macroscopic randomness", which corresponds to a Lyapunov timescale (i.e., λ^{-1}) of about 135 ($\approx 800 \times 0.1681$) for our three-body system. Figure 4.3 compares the convergent orbits for $d\mathbf{r}_1 = +10^{-60}$ and $d\mathbf{r}_1 = -10^{-60}$. The orbits are almost the same over the time interval $t \in [0, T^*]$, but increasingly deviate from each other when $t > T^* \approx 800$. Note that for $d\mathbf{r}_1 = +10^{-60}$, Body-1 escapes alone whereas Body-2 and Body-3 both move together increasingly far away from Body-1 to form a binary-body system. However, for $d\mathbf{r}_1 = -10^{-60}$, it is Body-3 that escapes alone while Body-1 and Body-2 move together further and further away from Body-3 and finally form a binary-body system. These results

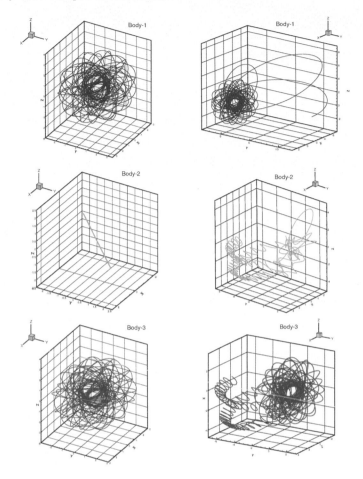

FIGURE 4.2

Comparison between convergent trajectories of the three-body system with three equal masses for initial conditions given by (4.20) and (4.21) when $d\mathbf{r}_1 = (0,0,0)$ (left) and $d\mathbf{r}_1 = 10^{-60} (1,1,1)$ (right). Upper plot: Body-1; middle plot: Body-2; and lower plot: Body-3.

all illustrate that macroscopic randomness arises from micro-level physical uncertainty of the initial condition $d\mathbf{r}_1$, with symmetry-breaking of the triple system and the escape of one of the three bodies exhibiting sensitivity to the micro-level physical uncertainty. For details, please refer to Liao [40].

Liao & Li [42] later considered 10,000 cases with different initial conditions (4.20) and (4.21) obtained by varying $d\mathbf{r}_1$ according to a Gaussian normal

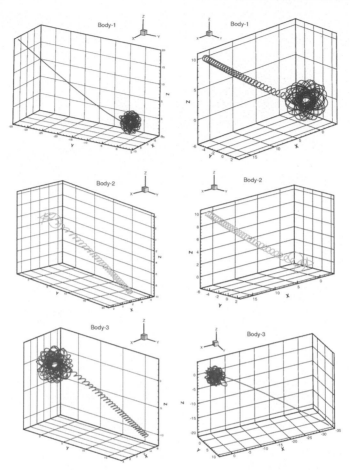

FIGURE 4.3

Comparison between convergent trajectories of the three-body system with three equal masses for initial conditions given by (4.20) and (4.21) when $d\mathbf{r}_1 = +10^{-60}(1, 0, 0)$ (left) and $d\mathbf{r}_1 = -10^{-60}(1, 0, 0)$ (right). Upper plot: Body-1; middle plot: Body-2; and lower plot: Body-3.

distribution with zero mean and standard deviation σ_0, such that

$$\langle d\mathbf{r}_1 \rangle = 0, \qquad \sigma_0 = \sqrt{\langle |d\mathbf{r}_1|^2 \rangle}, \tag{4.24}$$

where $\langle x \rangle$ denotes an operator that takes the average of x. Without loss of generality, let us first consider simulations when $\sigma_0 = 10^{-60}$. All the resulting chaotic three-body systems are found to have the same maximum Lyapunov exponent, i.e., $\lambda = 0.1681$. For each case of randomly generated micro-level physical uncertainty $d\mathbf{r}_1$, CNS invariably predicts a *convergent* chaotic orbit of the corresponding three-body system over a long time interval $t \in [0, 1000]$. It is found that *all* orbits exhibit no distinct deviation from each other during $t \in [0, T^*]$, where $T^* \approx 800$, but thereafter macroscopic randomness becomes increasingly evident, and includes symmetry-breaking, the escape of a single body (that leads to the disruption of the three-body system), etc., as shown in the left plots of Figure 4.4, which contain all the 10,000 convergent orbits given by CNS in the (x, y) plane at times $t = 810, 820$ and 830.

Liao & Li [42] further considered a different case for which $\sigma_0 = 3 \times 10^{-60}$, corresponding to a Gaussian normal distribution of $d\mathbf{r}_1$ with zero mean but a three times larger standard deviation. Using CNS, they obtained 10,000 *convergent* orbits in the time interval $t \in [0, 1000]$. It is very interesting that, at certain times such as $t = 810, 820$, and 830, the macroscopic randomness (indicated by for example the probability distribution) of the orbits is found to be sensitive to the standard deviation σ_0 of the Gaussian normal distribution of $d\mathbf{r}_1$, corresponding to micro-level uncertainty in the initial condition (4.20), as shown in Figures 4.4 and 4.5.

The foregoing illustrates that macroscopic randomness of the three-body system considered herein arises from micro-level physical uncertainty. In other words, micro-level physical uncertainty appears to be the origin of macroscopic randomness in a three-body system! This provides us with a theoretical evidence for the origin of macroscopic randomness.

4.1.3 Self-excited Escape and Symmetry-Breaking

What is the physical meaning of dimensionless tiny uncertainty $d\mathbf{r}_1$ of the initial position?

It is widely accepted that microscopic phenomena are essentially uncertain/random. The so-called Planck length

$$l_P = \sqrt{\frac{\hbar\, G}{c^3}} \approx 1.616252(81) \times 10^{-35} \quad (\text{m}) \qquad (4.25)$$

is the length scale at which quantum mechanics, gravity and relativity all interact very strongly, where $\hbar$ is the reduced Planck's constant, c is the speed of light in a vacuum, and G is the gravitational constant. According to string theory [76], the Planck length is the order of magnitude of oscillating strings that form elementary particles, and *a shorter length does not make physical sense*. Notably, in some forms of quantum gravity, it becomes *impossible* to determine the difference between two locations *less* than one Planck length apart. Therefore, at the level of the Planck length, the position of a body is *intrinsically* uncertain. This kind of microscopic physical uncertainty is *inherent* and has nothing to do with the Heisenberg uncertainty principle [77]; in other words, it is *objective*. Therefore, it is reasonable to assume that, from a physical standpoint, micro-level inherent fluctuations in the position of a

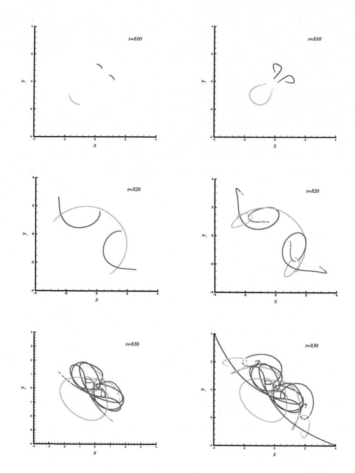

FIGURE 4.4

Comparison between the locational distributions of Body-1 (red points), Body-2 (green points) and Body-3 (blue points) in the (x, y) plane at different times for initial conditions given by (4.20), (4.21) and (4.24) when $\sigma_0 = 10^{-60}$ (left) and $\sigma_0 = 3 \times 10^{-60}$ (right). Results are based on 10000 convergent orbits during $t \in [0, 1000]$ given by CNS. Upper plot: $t = 810$; middle plot: $t = 820$; and lower plot: $t = 830$.

body shorter than the Planck length l_p are essentially uncertain and/or random. The foregoing uncertainty and randomness have mathematical meaning, but no physical meaning and so are treated as negligible from the physical perspective.

To render the Planck length $l_p \approx 1.62 \times 10^{-35}$ (meter) dimensionless, we select the diameter of the Milky Way Galaxy as the characteristic length, i.e.,

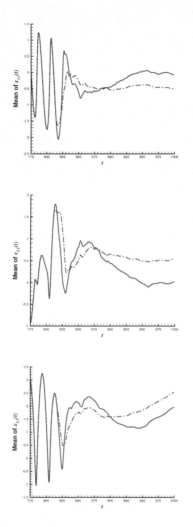

FIGURE 4.5
Time histories of mean location $(\overline{x}_{1,1}, \overline{x}_{2,1}, \overline{x}_{3,1})$ of Body-1 for $\sigma_0 = 10^{-60}$ and $\sigma_0 = 3 \times 10^{-60}$. The results are based on 10000 convergent orbits during the time interval $t \in [0, 1000]$ given by CNS. Solid line: $\sigma_0 = 10^{-60}$; dashed line: $\sigma_0 = 3 \times 10^{-60}$. Upper plot: mean of $x_{1,1}(t)$; middle plot: mean of $x_{2,1}(t)$; lower plot: mean of $x_{3,1}(t)$.

$d_G \approx 10^5$ light year $\approx 9 \times 10^{20}$ meter. Hence, the dimensionless Planck length is $l'_p = l_p/d_G \approx 1.8 \times 10^{-56}$, which is rather a small number. Thus, two points with dimensionless separation distance shorter than the dimensionless Planck length l'_p have *no* separate physical meaning: in other words, they are the *same* points in physics!

According to (4.20), for $d\mathbf{r}_1 = \pm 10^{-60}(1,0,0)$, we have

$$|d\mathbf{r}_1| = |\mathbf{r}_1 - (0,0,-1)| = 10^{-60} \ll l'_p. \tag{4.26}$$

Similarly, for $d\mathbf{r}_1 = 10^{-60}(1,1,1)$, we have

$$|d\mathbf{r}_1| = |\mathbf{r}_1 - (0,0,-1)| = \sqrt{3} \times 10^{-60} \ll l'_p. \tag{4.27}$$

In these cases, $|d\mathbf{r}_1|$ is much less than the dimensionless Planck length $l'_p = 1.8 \times 10^{-56}$, say, $|d\mathbf{r}_1| \ll l'_p$. Therefore, *from a physical viewpoint*, when $d\mathbf{r}_1 = 10^{-60}(1,1,1)$ or $d\mathbf{r}_1 = \pm 10^{-60}(1,0,0)$, the corresponding initial conditions are exactly the *same* as $d\mathbf{r}_1 = 0$.

However, as shown in Figures 4.2 and 4.3, these physically *same* initial conditions lead to three quite *different* orbits for $t > T^*$, including distinct macroscopic randomness such as symmetry-breaking of the system, the escape of a body, and so on, where $T^* \approx 800$ is the so-called "propagation time of microscopic uncertainty to macroscopic randmonness". It should be emphasized that this symmetry-breaking of the system and the escape of a body happen *without* any external disturbance forcing the system! Thus, symmetry-breaking of the three-body system and the escape of a body are *self-excited*, say, *without* any external disturbance. In other words, something happens due to nothing. We call these phenomena "self-excited symmetry-breaking" and "self-excited escape".

In addition, as shown in Figure 4.4, *all* orbits of the three-body system under the initial conditions (4.20), (4.21) and (4.24) obtained for the cases of $\sigma_0 = 10^{-60}$ and $\sigma_0 = 3 \times 10^{-60}$ have the *same* initial condition in physics, given that $|d\mathbf{r}_1| \ll l'_p$ holds in all cases! However, as shown in Figures 4.4 and 4.5, certain statistics of the three-body system, such as the probability of position distribution, the mean position of each body and so on, exhibit distinct differences when $t > T^* \approx 800$, where T^* is the propagation time of microscopic uncertainty to macroscopic randomness, even though the initial conditions are exactly the *same* in physics! In other words, the *physically* **same** initial conditions might lead to *statistically* **different** macroscopic results! This is very surprising! It illustrates that, statistics instability could occur in a chaotic system even *in the absence of* any external disturbance. For more details about the statistics instability of chaotic system, please refer to Chapter 5 of this book.

The foregoing illustrate that, due to the butterfly effect, micro-level physical uncertainty could transmit to macroscopic randomness. Here CNS provides us with a theoretical evidence that micro-level physical uncertainty might be the origin of macroscopic randomness. Our example demonstrates that CNS facilitates accurate investigation of the propagation of micro-level uncertainty even in a chaotic system.

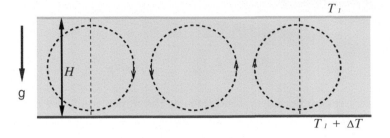

FIGURE 4.6
Schematic representation of two-dimensional Rayleigh-Bénard convective flow. Here, the incompressible fluid layer between two parallel free surfaces (separated by H) is heated from the lower which has a constant temperature difference $\Delta T > 0$ with respect to the upper whose temperature is set as T_1. The symbol g denotes gravitational acceleration.

4.2 Turbulent Rayleigh-Bénard Convection

It is of broad interest to understand how the evolution of non-equilibrium systems can be triggered by small external disturbances. A notable example is the origin of randomness in the transition from laminar to turbulent fluid flow, as identified in the seminal pipe flow experiment by Osborne Reynolds, which now forms a century old unresolved problem. Although there exist different hypotheses, it is widely believed that the randomness is "intrinsic". However, this remains an open hypothesis that is yet to be verified. By simulating the Rayleigh-Bénard convection system using CNS with negligible numerical noise below the level of thermal fluctuation, one can verify that turbulence can be self-excited from inherent thermal fluctuation, without any external disturbance, i.e., out of nothing.

4.2.1 Mathematical Model and the CNS Algorithm

Let us consider two-dimensional Rayleigh-Bénard convection (RBC) [78, 79]. As shown in Figure 4.6, an incompressible fluid layer between two horizontal free surfaces (separated by H) is heated by the lower free surface which has a prescribed constant temperature difference ΔT with respect to the upper free surface. Defining $\sqrt{g\alpha H \Delta T}$ as the characteristic velocity, where g is the acceleration due to gravity and α is the thermal expansion coefficient of the fluid, the non-dimensional governing equations with Boussinesq approximation read [78]

$$\frac{\partial}{\partial t}\nabla^2\psi + \frac{\partial\left(\psi,\nabla^2\psi\right)}{\partial(x,z)} - \frac{\partial\theta}{\partial x} - C_a\nabla^4\psi = 0, \tag{4.28}$$

$$\frac{\partial\theta}{\partial t} + \frac{\partial\left(\psi,\theta\right)}{\partial(x,z)} - \frac{\partial\psi}{\partial x} - C_b\nabla^2\theta = 0, \tag{4.29}$$

subject to free-slip boundary conditions at the upper and lower free surfaces

$$\frac{\partial\left(\psi,\nabla^2\psi\right)}{\partial(x,z)} = \frac{\partial\left(\psi,\theta\right)}{\partial(x,z)} = 0, \tag{4.30}$$

where (x,z) are horizontal and vertical spatial coordinates, t denotes time, ψ is the stream function, θ is the temperature departure from a linear variation background, and ∇^2 is the Laplace operator whereby definition $\nabla^4 = \nabla^2\nabla^2$,

$$\frac{\partial(a,b)}{\partial(x,z)} = \frac{\partial a}{\partial x}\frac{\partial b}{\partial z} - \frac{\partial b}{\partial x}\frac{\partial a}{\partial z}$$

is the Jacobian operator, and $C_a = \sqrt{Pr/Ra}$ and $C_b = 1/\sqrt{PrRa}$ are constants depending on the Rayleigh number $Ra = g\alpha H^3\Delta T/(\nu\kappa)$ and Prandtl number $Pr = \nu/\kappa$, in which ν is the kinematic viscosity and κ is the thermal diffusivity. Without loss of generality, let us consider the aspect ratio $\Gamma = L/H = 2\sqrt{2}$, Rayleigh number $Ra = 10^7$ and Prandtl number $Pr = 6.8$, corresponding to a linearly unstable situation of Rayleigh-Bénard convection.

Let us apply periodic boundary conditions in the horizontal direction, such that

$$\psi(x,z,t) = \psi(x+L,z,t), \qquad \theta(x,z,t) = \theta(x+L,z,t). \tag{4.31}$$

Thermal fluctuation at the micro-level plays an important role in hydrodynamic instability [80–82], and is used here as the initial condition. For water at 20°C (i.e., room temperature), the standard deviations of the initial temperature and velocity fields can be represented by statistical mechanics [83,84] as $\sigma_T = 10^{-10}$ and $\sigma_u = 10^{-9}$, respectively.

Following Saltzman [78], the stream function ψ and temperature departure θ are expressed in double Fourier expansion modes as

$$\psi(x,z,t) = \sum_{m=-\infty}^{+\infty}\sum_{n=-\infty}^{+\infty}\Psi_{m,n}(t)\exp\left[2\pi H\mathrm{i}\left(\frac{m}{L}x + \frac{n}{2H}z\right)\right]$$

$$\approx \sum_{m=-N_x}^{N_x}\sum_{n=-N_z}^{N_z}\Psi_{m,n}(t)\exp\left[2\pi H\mathrm{i}\left(\frac{m}{L}x + \frac{n}{2H}z\right)\right], \tag{4.32}$$

$$\theta(x,z,t) = \sum_{m=-\infty}^{+\infty}\sum_{n=-\infty}^{+\infty}\Theta_{m,n}(t)\exp\left[2\pi H\mathrm{i}\left(\frac{m}{L}x + \frac{n}{2H}z\right)\right]$$

$$\approx \sum_{m=-N_x}^{N_x}\sum_{n=-N_z}^{N_z}\Theta_{m,n}(t)\exp\left[2\pi H\mathrm{i}\left(\frac{m}{L}x + \frac{n}{2H}z\right)\right], \tag{4.33}$$

where m, n are the wave numbers in the x and z directions, $\Psi_{m,n}(t)$ and $\Theta_{m,n}(t)$ denote the amplitudes of the stream function and temperature components with wave numbers m and n, $\mathbf{i} = \sqrt{-1}$, respectively. For more details, please refer to Saltzman [78].

To ensure that the numerical simulation of spatiotemporal trajectory is convergent and thus reliable, background numerical noise (caused by round-off and truncation errors) must remain below the micro-level thermal fluctuation over a prescribed long time interval. This is impossible to achieve using conventional numerical algorithms with single (or double) precision floating-point arithmetic. Fortunately, CNS [38–40,43,54] makes it possible to overcome this challenge, as illustrated by Lin, Wang and Liao [44].

To ensure numerical accuracy, CNS is now used to simulate the evolution of micro-level thermal fluctuation. For our case of Rayleigh-Bénard convection with $Ra = 10^7$ and $Pr = 6.8$, Lin *et al.* [44] simulated thermal evolution using CNS with a 10th-order ($M = 10$) Taylor expansion in the temporal dimension, a time-step $\Delta t = 5 \times 10^{-3}$, and the double Fourier expansions with mode numbers $N_x = N_z = 128$ in the spatial x and z dimensions so as to reduce the truncation error to below a prescribed tiny level. Lin *et al.* [44] performed the computations in multiple-precision (MP) floating-point arithmetic with 100 significant digits ($N_s = 100$) so as to restrict the round-off error to below the prescribed tiny level.

To check convergence and reliability of their CNS result, Lin, Wang and Liao [44] computed another CNS result using the same Fourier expansion modes $N_x = N_z = 128$ in the spatial dimension but an even higher order ($M = 12$) of Taylor expansion in the temporal dimension with the same time-step Δt. The two CNS results were then compared at three probe points $(3L/4, H/10)$, $(3L/4, 2H/5)$, and $(3L/4, H/2)$, and it was found [44] that at all probe points the non-dimensional deviation between the two CNS results remained 10 orders of magnitude less than unity in the time interval $t \in [0, 50]$, and so were less than the thermal fluctuation under consideration. This confirmed that the CNS result given by the 10th-order Taylor expansion ($M = 10$) remained convergent and reliable through the time interval $t \in [0, 50]$.

4.2.2 From Thermal Fluctuation to Macroscopic Randomness

Although Rayleigh-Bénard convection is modelled theoretically as isolated from external disturbance, randomness at the microscopic level nevertheless exists because of molecular thermal fluctuation. Consider Case A and Case B with different initial temperature and velocity fields that are randomly generated as Gaussian white noise with the same temperature variance $\sigma_T = 10^{-10}$ and velocity variance $\sigma_u = 10^{-9}$. As shown in Figure 4.7(a), tiny differences in initial condition are negligibly small with respect to the background fields at the macroscopic level, and so the initial state can be regarded as the *same*

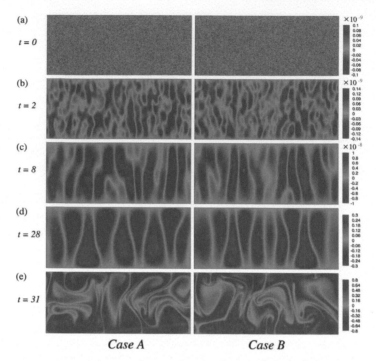

Case A   Case B

FIGURE 4.7

Evolution of the temperature departure from a linear variation background θ at a Rayleigh number $Ra = 10^7$ using double Fourier expansion modes $N_x = N_z = 127$. Cases A and B have different initial micro-level randomness caused by thermal fluctuation, generated by the same variance of temperature $\sigma_T = 10^{-10}$ and velocity $\sigma_u = 10^{-9}$. Contour plots are at times: (a) $t = 0$; (b) $t = 2$; (c) $t = 8$; (d) $t = 28$; and (e) $t = 31$.

from a physical viewpoint. However, the field structures and scales evolve rapidly as the time increases. Large-scale patterns appear even at a very early stage, as shown in Figure 4.7(b) for $t = 2$, although their magnitudes are still insensibly small. Later, the large-scale structures become increasingly distinct, as can be seen in Figure 4.7(c) at $t = 8$. Interestingly, these intermediate structures remain stable over a long time interval up to $t = 28$, as shown in Figure 4.7(d). At a critical point once the field is too energetic to be stable, these large-scale structures disintegrate abruptly, leading to the turbulent state visible in Figure 4.7(e) at $t = 31$. Note that the two flow structures in Figure 4.7(e) are completely different, and their deviation must originate from their different initial microscopic randomness due to thermal fluctuation. For details, please refer to Li, Wang, and Liao [44].

It should be emphasized that CNS can provide convergent, reliable simulations over a prescribed long interval of time while ensuring that numerical inaccuracy remains much less than the physical uncertainty, whereas direct numerical simulation (DNS) [16, 17] fails due to the butterfly effect. Therefore, *for the first time*, the CNS provides us with a theoretical evidence that turbulence in the Reyleigh-Bénard convection can be "self-excited" or simply arise "out of nothing" [85]. In other words, for the first time, this CNS result [44] provides us with a theoretical evidence that the origin of randomness in turbulence is indeed intrinsic!

In Chapter 6 of this book, the Reyleigh-Bénard convection is further used as an example to illustrate that the numerical noises as tiny artificial stochastic disturbances could lead to large-scale deviations of simulations not only in spatiotemporal trajectories but also even in statistics. For details, please also refer to Qin and Liao [55].

4.3 Origin of Macroscopic Randomness

Nature is full of randomness and uncertainty. It is well-known that the microscopic world is uncertain but can be mostly described by probability. But what is the origin of macroscopic randomness? Moreover, any random disturbances at a macro-level could lead to some macroscopic randomness. But what is the relationship between micro-level uncertainty and macroscopic randomness?

In Section 4.1, using CNS as a powerful tool to obtain convergent, reliable trajectories of chaos over a prescribed long time interval, we observed that symmetry-breaking and body-escape phenomena can occur *randomly* in a chaotic three-body system *without* any external disturbances, even if the physically *same* initial condition is used. It should of course be emphasized that, in the frame of CNS, numerical noise is several orders of magnitude smaller than physical micro-level uncertainty, so that numerical noise as a kind of artefact can be entirely neglected. In Section 4.1, we have seen that CNS provides us with a rigorous theoretical evidence that macroscopic randomness such as symmetry-breaking and body-escape in the chaotic three-body system can be *self-excited* and arise *out of nothing*.

In Section 4.2, we saw using CNS that even *intrinsic* micro-level thermal fluctuations (set as the initial condition) can lead to sharp deviations in solution trajectories for two-dimensional Rayleigh-Bénard convection (RBC) governed by the Navier–Stokes equations. It should be emphasized that, by means of CNS, numerical noise δ' as a kind of artefact is of much lower order of magnitude than the thermal fluctuations in temperature and velocity and so can be neglected compared to the true physical solution $\mathcal{P}$. Therefore, CNS provides us, *for the first time*, a rigorous theoretical evidence that random turbulence in Rayleigh-Bénard convection can be *self-excited*, i.e., occur *without* any external disturbances. In other words, the origin of macroscopic randomness in turbulence is essentially intrinsic!

In conclusion, the foregoing findings suggest that physical micro-level uncertainty might be the origin of macroscopic randomness, and that macroscopic randomness could even be *self-excited*, i.e., arise *without* any external disturbances!

Note that it is CNS that provides us with the capability to accurately investigate the propagation of micro-level physical uncertainty in chaotic systems over a prescribed long time interval. Indeed, CNS opens a door for us to enter a "clean" world of numerical simulations for chaos and turbulence, which should be closer to physical truth.

5

Ultra-chaos: A Higher Disorder Than Normal-chaos

In 1890, Poincaré [1] discovered that, for some dynamic systems, a small disturbance in initial condition can lead to huge deviation of trajectory after a sufficiently long time interval. Such dynamic systems are called chaotic systems, and such trajectory instability is called sensitivity dependence on initial condition (SDIC). In 1963, Lorenz [2] rediscovered SDIC which was later given the popular name "butterfly-effect". It is well-known that, due to SDIC, any environmental and/or artificial small disturbances of a chaotic system evolve to become exponentially enlarged to a macroscopic level. In 1989, Lorenz [9] further discovered that trajectories of chaotic systems also exhibit sensitivity dependence on artificial factors (SDAF) such as numerical algorithm, time-step, etc.

Note that all numerical simulations contain truncation and round-off errors, i.e., background numerical noise, which can be regarded as artificial disturbance. Due to the butterfly-effect, numerical noise as an artificial disturbance of a chaotic system enlarges exponentially, such that

$$\mathcal{E}(t) = \mathcal{E}_0 \exp(\kappa\, t), \quad t \in [0, T_c], \tag{5.1}$$

where $\mathcal{E}(t)$ is the level on average of the evolving deviation in the numerical trajectory from its true physical solution $\mathcal{P}$, $\mathcal{E}_0$ denotes the level of background numerical noise, and $\kappa > 0$ is the "noise-growth exponent", T_c is the so-called "critical predictable time" that is a key concept of CNS. Thus, in practice, a computer-generated numerical simulation $\mathcal{S}'$ of chaos quickly becomes a mixture $\mathcal{P} + \delta'$ of the true physical solution $\mathcal{P}$ and false numerical noise δ', which are mostly of the same order of magnitude, say, $\delta' \sim \mathcal{P} \sim \mathcal{S}'$, so that the numerical simulation $\mathcal{S}'$ becomes badly polluted by false numerical noise δ'. Unfortunately, there is *no* method presently available by which to separate the true physical solution $\mathcal{P}$ from the mixture $\mathcal{S}' = \mathcal{P} + \delta'$. Nevertheless, in practice, all statistics of chaotic systems are calculated using this kind of mixture $\mathcal{S}' = \mathcal{P} + \delta'$, which is based on the following assumed **hypothesis** that a statistic obtained from the mixture $\mathcal{S}' = \mathcal{P} + \delta'$ is invariably the same as the corresponding statistic of the true physical solution $\mathcal{P}$, say,

$$\langle \mathcal{P} + \delta' \rangle = \langle \mathcal{P} \rangle, \quad \text{when } \delta' \sim \mathcal{P} \text{ mostly}, \tag{5.2}$$

DOI: 10.1201/9781003299622-5

where $\langle \rangle$ is a statistical operator. In essence, hypothesis (5.2) implies that statistics of a chaotic system are *stable* to small disturbances. In fact, the direct numerical simulation (DNS) [16,17], which has no idea about the so-called "critical predictable time" T_c, is exactly based on this hypothesis.

Does hypothesis (5.2) indeed *always* hold in general? This is a fundamental problem of chaos dynamics. Unfortunately, hypothesis (5.2) has *not* been proved to hold for general cases, even though it has been widely utilized in a huge number of publications. This is a very serious problem.

Clean Numerical Simulation (CNS) [38–40,43,54] provides us with a means by which to test the hypothesis (5.2) because $\langle \mathcal{P} + \delta' \rangle$ can be easily obtained using conventional algorithms, and $\langle \mathcal{P} \rangle$ can be accurately determined by CNS, separately. The approach works because the false numerical noise δ' of a chaotic computer-generated simulation $\mathcal{S}$ given by CNS is several orders of magnitude lower than the true physical solution $\mathcal{P}$, i.e., $|\delta'| \ll |\mathcal{P}|$, over a prescribed time duration $t \in [0, T_c]$, so that the simulation $\mathcal{S} = \mathcal{P} + \delta'$ as a mixture of $\mathcal{P}$ and δ' is nevertheless very close to the true physical solution $\mathcal{P}$, say, $\mathcal{S} \approx \mathcal{P}$.

In this chapter, following hypothesis (5.2), we further classify chaos into normal-chaos whose statistics are stable and ultra-chaos whose statistics are unstable. Several examples are used to illustrate the widespread existence of ultra-chaos. It is concluded that the new concept of ultra-chaos could greatly enrich our understanding of chaos dynamics, and provide insight into limitations of the current paradigm of the scientific method.

5.1   A New Classification: Normal-chaos and Ultra-chaos

The trajectories of a chaotic system are unstable to small disturbances due to sensitivity dependence on initial condition (SDIC) [1], i.e., the butterfly-effect [2]. It has been long established that a chaotic system has at least one positive Lyapunov exponent [5,7,8]. A chaotic system with two positive Lyapunov exponents is termed hyper-chaos [63,86].

Due to the butterfly-effect, numerical noise increases exponentially in a chaotic system. Traditionally, there do *not* exist any methods by which to separate the true physical solution $\mathcal{P}$ from a chaotic numerical simulation $\mathcal{S}'$ given by conventional algorithms, which is a kind of mixture, say, $\mathcal{S}' = \mathcal{P} + \delta'$, where δ' denotes numerical noise that is mostly of the same order of magnitude as $\mathcal{P}$, say, $\delta' \sim \mathcal{P} \sim \mathcal{S}'$. However, CNS [38–40, 42, 43] can provide a convergent, reliable numerical simulation of chaos over a sufficiently long time interval $[0, T_c]$, during which false numerical noise δ' is very much smaller than the true physical solution $\mathcal{P}$, i.e., $|\delta'| \ll |\mathcal{P}|$, and so may be neglected. Hence, the CNS result $\mathcal{S}$ as a mixture $\mathcal{S} = \mathcal{P} + \delta'$ provides a very good approximation to the true physical solution $\mathcal{P}$ during $t \in [0, T_c]$, i.e.,

$\mathcal{S} \approx \mathcal{P}$, and so can be used as a *benchmark* solution whereby $\langle \mathcal{P} \rangle \approx \langle \mathcal{S} \rangle$. In the foregoing, T_c is the so-called "critical predictable time" that should be at least long enough for the statistics to be properly extractable. Comparing the CNS benchmark solution with corresponding solutions given by conventional algorithms in single (or double) precision floating-point arithmetic (where the numerical noise is mostly of the same order of magnitude as the true physical solution), one can study the influence of numerical noise on the statistics of a chaotic system by testing hypothesis (5.2). So, CNS provides with a practical way to accurately investigate the stability of statistics (to small disturbances) of dynamical systems.

Without loss of generality, let us first consider the Lorenz equation

$$\begin{cases} \dot{x} = \sigma\,(y - x), \\ \dot{y} = R\,x - y - x\,z, \\ \dot{z} = x\,y + B\,z, \end{cases} \tag{5.3}$$

for a chaotic case in which $\sigma = 10$, $R = 28$ and $B = -8/3$, subject to the initial condition

$$x(0) = -15.8, \quad y(0) = -17.48, \quad z(0) = 35.64. \tag{5.4}$$

Liao & Wang [41] gained a convergent, reliable chaotic simulation of the Lorenz equations (5.3) and (5.4) over a long time interval $t \in [0, 10000]$ using a parallel CNS algorithm. This provided a benchmark solution (marked as CNS) that is very close to the true physical solution $\mathcal{P}$ and so can provide accurate statistics $\langle \mathcal{P} \rangle$ for testing the hypothesis (5.2). For comparison purposes, the same equations (5.3) and (5.4) are solved during $t \in [0, 10000]$ using a conventional fourth-order Runge-Kutta method in single-precision (denoted RKwS) or double-precision (denoted RKwD) for different values of time-step Δt, corresponding to different levels of numerical noise as artificial disturbances. Note that trajectories given by the fourth-order Runge-Kutta algorithm in single/double precision for different time-steps Δt greatly deviate from each other after $t > T_c$, where $T_c < 32$ for all cases under consideration[*]. In other words, chaotic trajectories of the Lorenz system are *unstable* to small disturbances, i.e., numerical noise. However, the statistics given by the probability density function (PDF) and autocorrelation functions (ACFs) of $x(t), y(t)$, and $z(t)$, based on the quite different trajectories during $t \in [0, 10000]$ given by the conventional Runge-Kutta method, agree quite well with the corresponding statistics based on the CNS benchmark solution given by Liao & Wang [41], as shown in Figure 5.1. Therefore, the statistics of the *unstable* trajectories of the Lorenz equations (5.3) and (5.4) are *stable*! It should be emphasized that it is CNS that provides us, for the first time, with *rigorous* evidence that

---

[*]In fact, for each numerical algorithm, there exists a "critical predictable time" T_c before which false numerical noise δ' in the corresponding simulation is negligible compared to the true physical solution $\mathcal{P}$. For the Lorenz equations (5.3) and (5.4), the value of T_c is very small when the conventional Runge-Kutta method is implemented.

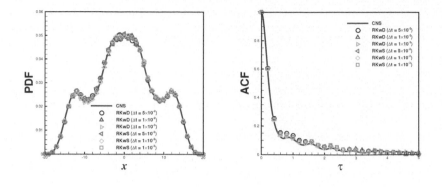

FIGURE 5.1

Influence of numerical noise on the statistics of normal-chaos. (a): probability density function (PDF) of $x(t)$; (b): autocorrelation function (ACF) of $x(t)$, i.e., $\langle x(t)x(x+\tau)\rangle$. The statistical results are based on simulations of a chaotic trajectory $x(t)$ over the time interval $0 \le t \le 10000$ governed by the Lorenz equations (5.3) and (5.4), given by CNS (red line) and Runge-Kutta algorithms (symbols) in double-precision (RKwD) or single-precision (RKwS) for different time-steps Δt. Qualitatively, the same findings are obtained for $y(t)$ and $z(t)$.

hypothesis (5.2) indeed holds for the Lorenz model. Thus, if one is only interested in the statistics of the Lorenz model, the conventional Runge-Kutta method in single (or double) precision floating-point arithmetic is sufficient. This is both fantastic and wonderful! Such chaos is called "normal-chaos", whose statistics are *stable* to small disturbances. Note that hypothesis (5.2) holds for a normal-chaotic system.

In fact, the statistics of many chaotic systems are stable, i.e., insensitive to small disturbances, and thus belong to normal-chaos, including hyper-chaos [63, 87, 88] which has two positive Lyapunov exponents, and the Lorenz-84 climate model [89]. For details, please refer to Liao and Qin [54].

Similarly, it has been found [49–51,55,56] that the statistics of some chaotic systems are *unstable*, i.e., rather sensitive to small disturbances. These systems are called ultra-chaotic [54]. This is very interesting! Obviously, ultra-chaos whose statistics are unstable is essentially different from normal-chaos whose statistics are stable. Note that hypothesis (5.2) does *not* hold for ultra-chaos, say

$$\langle \mathcal{P} + \delta' \rangle \ne \langle \mathcal{P} \rangle, \quad \text{when } \delta' \sim \mathcal{P} \text{ mostly}, \tag{5.5}$$

where $\langle \rangle$ denotes a statistical operator, $\mathcal{P}$ denotes the true physical solution, δ' is numerical noise, and $\mathcal{S}' = \mathcal{P} + \delta'$ is a numerical simulation (comprising a mixture of $\mathcal{P}$ and δ') gained by conventional algorithms using single (or

TABLE 5.1
Stability of trajectory and statistics of different dynamic systems

Type	Trajectory	Statistics	Measure of Disorder
Non-chaos	Stable	Stable	0
Normal-chaos	Unstable	Stable	1
Ultra-chaos	Unstable	Unstable	2

double) precision arithmetic. It should be emphasized that it is CNS that provides us, for the first time, with such a rigorous evidence that hypothesis (5.2) does *not* hold in some cases. This again illustrates the novelty of CNS.

Table 5.1 presents the new classification of dynamic systems according to the stability of their trajectories and statistics. A dynamic system is non-chaotic when both its trajectory and statistics are stable to small disturbances, normal-chaotic when its trajectories are unstable but its statistics are stable to small disturbances, and ultra-chaotic when both its trajectories and statistics are unstable to small disturbances. In practice, (5.5) provides us with a criterion by which to distinguish ultra-chaos from normal-chaos.

Let $\mathcal{I}_m$ denote a binary measure of instability, defined such that $\mathcal{I}_m = 0$ refers to stability and $\mathcal{I}_m = 1$ refers to instability. Let $\mathcal{I}_m^T$ and $\mathcal{I}_m^S$ denote the instability measure of trajectory and statistics. We therefore define the "disorder measure" as

$$\mathcal{D}_m = \mathcal{I}_m^T + \mathcal{I}_m^S. \tag{5.6}$$

As shown in Table 5.1, the disorder measure is 0 for non-chaos, 1 for normal-chaos, and 2 for ultra-chaos. Thus, according to (5.6), ultra-chaos has higher disorder than normal-chaos.

Note that hyper-chaos [63, 87, 88], which is defined as having two positive Lyapunov exponents, in fact belongs to normal-chaos confirming that ultra-chaos is indeed a new concept. Section 5.2 examines several examples of ultra-chaos, and Section 5.3 discusses its scientific meaning.

5.2   Examples of Ultra-chaotic Systems

5.2.1   Sine-Gordon Equation

Consider the damped driven sine-Gordon equation [73–75] with Gaussian white noise $\epsilon_0(x, t)$:

$$u_{tt} = u_{xx} - \sin(u) - \alpha u_t + \Gamma \sin(\omega t - \lambda x) + \epsilon_0(x, t), \tag{5.7}$$

subject to zero initial condition

$$u(x, 0) = 0, \qquad u_t(x, 0) = 0, \tag{5.8}$$

and periodic boundary condition

$$u(x + l, t) = u(x, t), \tag{5.9}$$

where the subscript denotes the partial derivative, x and t are spatial and temporal variables, α and Γ are physical parameters related to damped friction and external force, ω and $\lambda = 2\pi/l$ are temporal and spatial frequencies, and l is the total length. Following Chacón *et al.* [74], let us consider cases involving the following three fixed parameters

$$\alpha = \frac{1}{10}, \; l = 500, \lambda = \frac{2\pi}{l}, \tag{5.10}$$

and two variable parameters $\omega \in [0, 1.4]$ and $\Gamma \in [0, 1.4]$.

Here, Gaussian white noise $\epsilon_0(x, t)$ denotes environmental disturbance. Let σ_n denote the standard deviation of $\epsilon_0(x, t)$. Without loss of generality, let us consider three cases: $\sigma_n = 0$, $\sigma_n = 10^{-18}$, and $\sigma_n = 10^{-20}$. By definition, no disturbance exists when $\sigma_n = 0$.

For $\sigma_n = 0$, and given ω and Γ, one can use CNS to gain the convergent trajectory in the same way as mentioned by Liao and Qin [54]. In this case, the CNS result lies close to the true physical solution $\mathcal{P}$ and thus can be used as a benchmark solution to investigate the influence of the small disturbance $\epsilon_0(x, t)$ by comparison against the solutions obtained for $\sigma_n \neq 0$. When external disturbance is present, i.e., $\sigma_n \neq 0$, we first calculate $u'(x, t + \Delta t)$ (where Δt denotes the time-step) by CNS, then add Gaussian white noise ϵ_0, so as to finally simulate

$$u(x, t + \Delta t) = u'(x, t + \Delta t) + \epsilon_0. \tag{5.11}$$

Thereafter, criterion (5.2) is used to check whether a simulation relates to normal-chaos or ultra-chaos. In practice, we classify the system as being ultra-chaotic if the difference between its statistics under small disturbances and the corresponding statistics without disturbances given by the CNS benchmark solution is larger than 5%.

Here, in the frame of CNS, the spatial domain is discretized into $N = 2^{16} = 65536$ equidistant points, and the governing equations solved in multiple-precision floating-point arithmetic with $N_s = 230$ significant digits used for all variables and parameters, and a variable time-step applied with allowable tolerance $tol = 10^{-230}$. In this way, the corresponding background numerical noise is much lower than the tiny external disturbance so that we obtain convergent, reliable chaotic simulations throughout the whole spatial domain over the time interval $t \in [0, 3600]$, which is *sufficiently* long for statistical analysis. For more details, please refer to Liao and Qin [54].

As reported by Liao and Qin [54], the system governed by the damped driven sine-Gordon equations (5.7)–(5.10) for $\omega \in [0, 1.4]$ and $\Gamma \in [0, 1.4]$ can be classified into three different types of dynamic systems; namely, non-chaos, normal-chaos and ultra-chaos, as shown in Figure 5.2. This is a good example to illustrate the pervasive existence of ultra-chaos. Note that, when $\Gamma = 3/5$

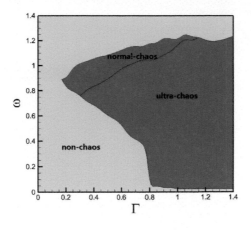

FIGURE 5.2
Classification of solutions of the damped driven sine-Gordon equations (5.7)–(5.10) for $\omega \in [0, 1.4]$ and $\Gamma \in [0, 1.4]$. Red domain: ultra-chaos; blue domain: normal-chaos; gray domain: non-chaos.

and as ω increases from 0 to 1.4, the state of the system transfers from non-chaos to ultra-chaos, then from ultra-chaos to normal-chaos, and finally from normal-chaos to non-chaos.

Without loss of generality, let us first consider normal-chaos for $\Gamma = 3/5$ and $\omega = 0.98$. Its statistics, including the probability density function (PDF) (given by the tempo-spatial average) and the total spectral energy $E_s(t)$ of $u(x, t)$, are *stable* (i.e., insensitive) to small disturbances, as shown in Figure 5.3. The total spectral energy is given by

$$E_s(t) = \sum_{k=-\infty}^{k=+\infty} \left| c_k(t) \right|^2, \tag{5.12}$$

in which c_k is the time-dependent amplitude of the Fourier series

$$u(x, t) = \sum_{k=-\infty}^{+\infty} c_k(t) \exp(k \lambda x \, \mathbf{i}), \tag{5.13}$$

and $\mathbf{i} = \sqrt{-1}$. In this case, the PDFs obtained from numerical simulations given by traditional algorithms in single (or double) precision floating-point arithmetic agree well with the PDF given by the CNS benchmark solution. So, from the viewpoint of numerical computation, it is easy to solve a case involving normal-chaos if one is interested in its statistics only.

Now, let us consider ultra-chaos for $\omega = 0.6$ and $\Gamma = 0.922$. Figure 5.4 shows the influence of small disturbances on the probability density functions

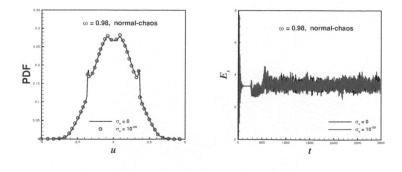

FIGURE 5.3

Influence of tiny disturbances on the probability density function (PDF) and the total spectral energy $E_s(t)$ of normal-chaos governed by the damped driven sine-Gordon equations (5.7)–(5.10) for $\omega = 0.98$ and $\Gamma = 3/5$ and two different levels of white noise: $\sigma_n = 0$ (black); and $\sigma_n = 10^{-20}$ (blue). Left plot: PDF; right plot: $E_s(t)$.

(PDFs) and total spectral energy $E_s(t)$ of the ultra-chaotic simulation $u(x, t)$ for the three levels of white noise, $\sigma_n = 0$, $\sigma_n = 10^{-18}$, and $\sigma_n = 10^{-20}$. Note that, the PDF is very sensitive to small disturbance: at $u = 0$, the relative error of the PDFs reaches 62% for $\sigma_n = 10^{-20}$ compared against the benchmark result given by CNS when $\sigma_n = 0$. The total spectral energy $E_s(t)$ is also very

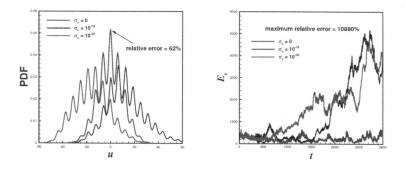

FIGURE 5.4

Influence of small disturbances on probability density functions (PDFs) and the total spectral energy $E_s(t)$ of ultra-chaos $u(x, t)$, governed by the damped driven sine-Gordon equations (5.7)–(5.10) for $\omega = 0.6$ and $\Gamma = 0.922$ and three different levels of white noise: $\sigma_n = 0$ (red); $\sigma_n = 10^{-18}$ (black); and $\sigma_n = 10^{-20}$ (blue). Left plot: PDF; right plot: $E_s(t)$.

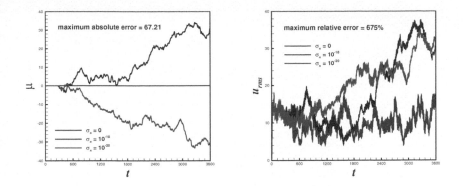

FIGURE 5.5
Influence of small disturbances on spatial statistics of ultra-chaos $u(x, t)$, governed by the damped driven sine-Gordon equations (5.7)–(5.10) for $\omega = 0.6$ and $\Gamma = 0.922$ and three different levels of white noise: $\sigma_n = 0$ (red), $\sigma_n = 10^{-18}$ (black); and $\sigma_n = 10^{-20}$ (blue). (a) Temporal profile of spatial mean $\mu(t)$; (b) Temporal profile of spatial root-mean-square (RMS) $u_{rms}(t)$.

sensitive to the level of small disturbances, with the maximum $E_s(t)$ exceeding 5000 for $\sigma_n = 10^{-18}$ and $\sigma_n = 10^{-20}$, whereas the maximum total spectral energy of the CNS benchmark solution remains below 600. Obviously, the total spectral energy $E_s(t)$ obtained for the tiny disturbances $\sigma_n = 10^{-18}$ and $\sigma_n = 10^{-20}$ exhibits huge deviation (maximum relative error of 10880%) with respect to the CNS benchmark solution without disturbance (i.e., for $\sigma_n = 0$). Comparing Figure 5.4 with Figure 5.3, we can see that the maximum total spectral energy $E_s(t)$ of ultra-chaos is more than 500 times larger than that of normal-chaos. This numerical experiment suggests that ultra-chaos might contain much larger energy than normal-chaos. This is reasonable because ultra-chaos is of higher disorder than normal-chaos.

Figure 5.5 shows the influence of small disturbances on the spatial mean $\mu(t)$ and spatial root-mean-square (RMS) $u_{rms}(t)$ of $u(x, t)$, expressed by the spatial average, for disturbance levels of $\sigma_n = 0$, $\sigma_n = 10^{-18}$, and $\sigma_n = 10^{-20}$. Note that these spatial statistics are very sensitive to the level of tiny disturbances. The maximum absolute error of $\mu(t)$ is 67.21 and the maximum relative error of $u_{rms}(t)$ is 675%. Note that, when there is no environmental noise (i.e., $\sigma_n = 0$), the spatial mean $\mu(t)$ is almost zero and the spatial root-mean-square $u_{rms}(t)$ is always less than 20. However, the tiny noise leads inevitably to a gradually increasing deviation in $\mu(t)$ from zero, which finally results in $u_{rms}(t) > 30$.

Figure 5.6 shows the impact of the small disturbances on the temporal statistics, i.e., the temporal mean $\mu(x)$ and the temporal root-mean-square

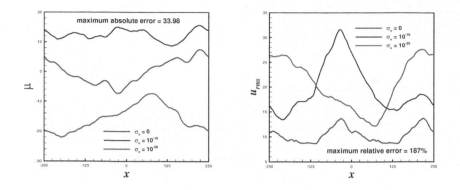

FIGURE 5.6
Influence of small disturbances on temporal statistics of ultra-chaos $u(x,t)$, governed by the damped driven sine-Gordon equations (5.7)–(5.10) for $\omega = 0.6$ and $\Gamma = 0.922$ and three different levels of white noise: $\sigma_n = 0$ (red); $\sigma_n = 10^{-18}$ (black); and $\sigma_n = 10^{-20}$ (blue). (a) Spatial profile of temporal mean $\mu(x)$; (b) Spatial profile of temporal root-mean-square (RMS) $u_{rms}(x)$.

(RMS) $u_{rms}(x)$, given by the temporal average, for $\sigma_n = 0$, $\sigma_n = 10^{-18}$ and $\sigma_n = 10^{-20}$. It is clear that the temporal statistics are also very sensitive to the level of the small disturbances. Here, the maximum absolute error of $\mu(x)$ is 33.98 and the maximum relative error of $u_{rms}(x)$ is 187%.

The foregoing illustrates that both normal-chaos and ultra-chaos widely exist in the system governed by the damped driven sine-Gordon equations (5.7)–(5.10). However, unlike normal-chaos, the statistics of ultra-chaos are unstable in that they are highly sensitive to very tiny disturbances. The findings verify once again that ultra-chaos is indeed of higher disorder than normal-chaos. For more details about the above-mentioned comparisons, please refer to Liao and Qin [54].

5.2.2   Arnold-Beltrami-Childress (ABC) Flow

The Arnold-Beltrami-Childress (ABC) flow has a velocity field given by

$$\mathbf{u}_{ABC} = \left(A \sin z + C \cos y, B \sin x + A \cos z, C \sin y + B \cos x\right) \qquad (5.14)$$

that describes the stationary flow of an incompressible fluid with periodic boundary conditions, where x, y, and z are Cartesian coordinates, and A, B, and C are constants. Arnold [90] was the first to discover that the velocity fields of ABC flows (5.14) form a class of steady-state solutions of the Euler equations, or the Navier–Stokes (NS) equations with an external force per unit mass. Note that the ABC flow possesses the property of Lagrangian chaos

[91–95] in that the trajectory of a fluid particle (observed from a Lagrangian viewpoint which follows the motion of a fluid tracer as it moves in space and time) deviates exponentially when subject to a small initial disturbance.

In the Lagrangian framework, the motion of a fluid particle in ABC flow (5.14) is governed by

$$\begin{cases} \dot{x}(t) = A\sin[z(t)] + C\cos[y(t)], \\ \dot{y}(t) = B\sin[x(t)] + A\cos[z(t)], \\ \dot{z}(t) = C\sin[y(t)] + B\cos[x(t)], \end{cases} \tag{5.15}$$

subject to the initial condition

$$(x(0), y(0), z(0)) = \mathbf{r}_0, \tag{5.16}$$

where $\mathbf{r}_0$ denotes the starting position of the fluid particle.

Without loss of generality, let us consider fixed $A = 1$ but varying B and C. Convergent trajectories of chaotic motion of a fluid particle in the ABC flow can be obtained by means of CNS over a time interval that is sufficiently long for statistical analysis to be meaningfully accurate. Assume that, due to a small disturbance, the starting position experiences a tiny deviation from $\mathbf{r}_0 = (x(0), y(0), z(0))$ so that $\mathbf{r}'_0 = \mathbf{r}_0 + (0, 0, 1) \times \delta$, where $\delta = |\mathbf{r}'_0 - \mathbf{r}_0|$ is a tiny constant. Note that $\mathbf{r}'_0 = \mathbf{r}_0$ when $\delta = 0$, corresponding to a case without disturbance. Again without loss of generality, let us consider two cases that correspond to rather small levels of disturbance: $\delta = 10^{-5}$ and $\delta = 10^{-10}$. In both cases, the convergent trajectory over the time interval $t \in [0, 10000]$ can be obtained by means of CNS using a 200th-order Taylor expansion, time step $\Delta t = 0.01$ and 500 digits ($N_s = 500$) multiple-precision (MP) floating point arithmetic. For more details about the CNS algorithms used here, please refer to Qin and Liao [57].

First, let us consider the trajectory of the fluid particle starting from $(0, 0, 0)$ in the ABC flow described by Eqs. (5.15) and (5.16). For $A = 1$, $B = 0.7$, and $C = 0.42$, the trajectory is chaotic with maximum Lyapunov exponent $\lambda_{max} = 0.01$. Although the three trajectories corresponding to $\delta = 0$, 10^{-5}, and 10^{-10} depart from each other quickly, their phase plots and probability density functions (PDFs) are almost the same, as shown in Figure 5.7. In other words, although the trajectory is unstable, its statistics are stable, i.e., insensitive to small disturbances. Therefore, when $A = 1$, $B = 0.7$, and $C = 0.42$, the trajectory of the fluid particle starting from $(0,0,0)$ behaves as normal-chaos.

Next, let us increase C from 0.42 a relatively small amount to 0.43, such that $\Delta C = 0.01$. The chaotic trajectory (with maximum Lyapunov exponent $\lambda_{max} = 0.06$) of a fluid particle starting from $(0,0,0)$ in ABC flow for $A = 1$, $B = 0.7$, and $C = 0.43$ behaves quite differently from the trajectory for $A = 1$, $B = 0.7$, and $C = 0.42$ mentioned above: its phase plots and probability density functions are very sensitive to the small disturbance, as shown in Figure 5.8,

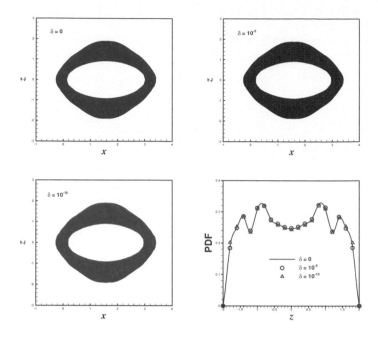

FIGURE 5.7

Influence of tiny disturbances on the phase plot $x - z$ and the probability density function (PDF) of normal-chaotic motion of a fluid particle in ABC flow (5.14) for $A = 1$, $B = 0.7$, and $C = 0.42$, plotted from the starting point (and two nearby points representing small disturbances) $\mathbf{r}'_0 = (0, 0, 0) + (0, 0, 1) \times \delta$. Upper-left: (x, z) phase plot obtained for $\delta = 0$; upper-right: (x, z) phase plot obtained for $\delta = 10^{-5}$; lower-left: (x, z) phase plot obtained for $\delta = 10^{-10}$; and lower-right: PDFs of $z(t)$. Red: $\delta = 0$; black: $\delta = 10^{-5}$; and blue: $\delta = 10^{-10}$.

where the PDF of $z(t)$ is based on its normalized value $-\pi \leq z' < +\pi$ via the periodic condition

$$z'(t) = z(t) + 2\pi\, n_z, \tag{5.17}$$

with n_z being an integer. Thus, the corresponding motion of the fluid particle starting from (0,0,0) for $A = 1$, $B = 0.7$, and $C = 0.43$ comprises ultra-chaos. Obviously, ultra-chaos involves a higher level of disorder than normal-chaos. This provides solid evidence that ultra-chaos can indeed exist in an ABC flow field (5.15) and (5.16).

Furthermore, let us consider the ensemble average of chaotic trajectories of fluid particles starting from one thousand nearby points located at

$$\mathbf{r}_0 = (0, 0, 0) + (0, 0, 1) \times \delta_i,$$

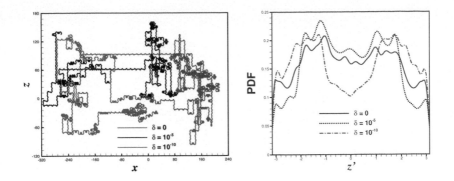

FIGURE 5.8

Influence of tiny disturbances on the phase plot $x - z$ and the probability density function (PDF) of ultra-chaotic motion of a fluid particle in ABC flow (5.14) for $A = 1$, $B = 0.7$, and $C = 0.43$, plotted from the starting point (and two nearby points representing small disturbances) $\mathbf{r}_0' = (0, 0, 0) + (0, 0, 1) \times \delta$. Left: (x, z) phase plot; right: PDFs of $z(t)$. Red: $\delta = 0$; black: $\delta = 10^{-5}$; and blue: $\delta = 10^{-10}$.

where δ_i ($i = 1, 2, 3, ..., 1000$) is specified by Gaussian white noise of standard deviation $\sigma_d = \sqrt{\langle \delta_i^2 \rangle}$ and zero mean, i.e., $\langle \delta_i \rangle = 0$. Here, $\langle \rangle$ denotes the ensemble average operator. Figure 5.9 shows that the PDF of the ensemble-averaged trajectory is stable to small disturbances for $A = 1$, $B = 0.7$, and $C = 0.42$, but is very sensitive (i.e., unstable) to the tiny standard deviations ($\sigma_d = 10^{-5}$ or 10^{-10}) for $A = 1$, $B = 0.7$, and $C = 0.43$. Thus, even from the ensemble-average viewpoint, the trajectory of a fluid particle starting from $(0,0,0)$ in ABC flow (5.15) and (5.16) behaves as normal-chaos when $A = 1$, $B = 0.7$, and $C = 0.42$, but is ultra-chaotic when $A = 1$, $B = 0.7$, and $C = 0.43$. This further confirms that an ultra-chaotic trajectory indeed has higher disorder than a normal-chaotic trajectory.

In a similar fashion, we can investigate the trajectory property of a fluid particle starting from $(0,0,0)$ for fixed $A = 1$ but varying B and C. Figure 5.10 shows that the trajectory might be non-chaos, normal-chaos, and ultra-chaos, indicating the wide prevalence of ultra-chaos in the ABC flow (5.15) and (5.16). For non-chaotic motion, both the trajectory and its statistics are *stable* to tiny disturbances. For normal-chaotic motion, although the trajectory is unstable, its phase plots and statistics are *stable* to tiny initial disturbances. However, for ultra-chaotic motion, both its trajectory and statistics are *unstable*. Note that, for normal-chaos, the fluid particle always moves in a restricted spatial domain, such as the example displayed in the phase plot $x - z$ of Figure 5.7. However, for ultra-chaos, the fluid particle

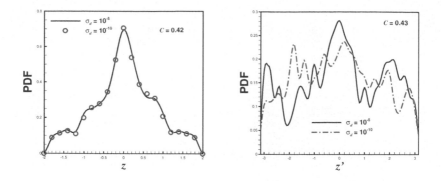

FIGURE 5.9

Influence of tiny disturbances on the PDF of the ensemble-averaged trajectory of a normal-chaotic (left) or an ultra-chaotic (right) fluid particle in ABC flow (5.14) for $A = 1$, $B = 0.7$, and either $C = 0.42$ (left) or $C = 0.43$ (right), from the starting points $\mathbf{r}_0 = (0,0,0) + (0,0,1) \times \delta_i$, $1 \le i \le 1000$, with the standard deviation $\sigma_d = \sqrt{\langle \delta_i^2 \rangle} = 10^{-5}$ (black) and $\sigma_d = 10^{-10}$ (blue), respectively. Left: PDF of $z(t)$ of normal-chaotic fluid particle when $C = 0.42$; right: PDF of normalized $z(t)$ of ultra-chaotic fluid particle when $C = 0.43$.

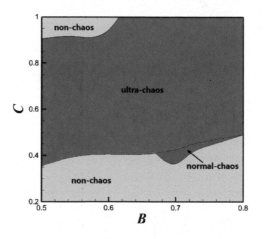

FIGURE 5.10

Classification of fluid particle trajectories in ABC flow (5.14) starting from $\mathbf{r}_0 = (0,0,0)$ for $A = 1$ and varying $B \in [0.5, 0.8]$ and $C \in [0.2, 1]$. Red: ultra-chaos; blue: normal-chaos; gray: non-chaos.

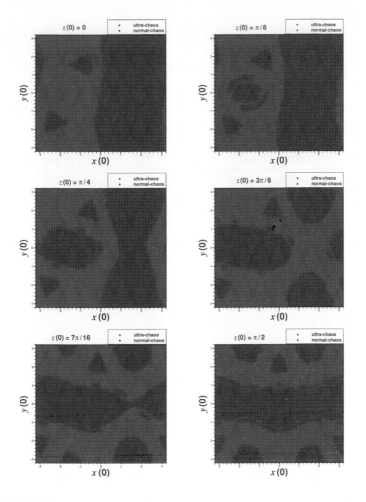

FIGURE 5.11

Classification of fluid particle trajectories starting from different points $\mathbf{r}_0 = (x(0), y(0), z(0))$ where $x(0) \in [-2\pi, 2\pi]$ and $y(0) \in [-2\pi, 2\pi]$ in ABC flow (5.14) for $A = 1$, $B = 0.7$, and $C = 0.43$. Upper-left: $z(0) = 0$; upper-right: $z(0) = \pi/8$; middle-left: $z(0) = \pi/4$; middle-right: $z(0) = 3\pi/8$; lower-left: $z(0) = 7\pi/16$; lower-right: $z(0) = \pi/2$. Red: ultra-chaos; blue: normal-chaos.

progressively departs from its starting position, again illustrating that ultra-chaos has higher disorder than normal-chaos.

Keeping $A = 1$, $B = 0.7$, and $C = 0.43$ fixed but using various starting positions $\mathbf{r}_0 = (x(0), y(0), z(0))$ in the ABC flow (5.15) and (5.16), where $-\pi \leq x(0), y(0), z(0) \leq +\pi$, it is found that both normal-chaos and ultra-chaos co-exist simultaneously, as shown in Figure 5.11. Statistics of the maximum

TABLE 5.2

Maximum Lyapunov exponents λ_{max} of normal-chaotic and ultra-chaotic trajectories of fluid particles in ABC flow (5.14) for $A = 1$, $B = 0.7$, and $C = 0.43$, commencing from various starting points $\mathbf{r}_0 = (x(0), y(0), z(0))$, where $-\pi \le x(0), y(0), z(0) \le +\pi$

	Normal-chaos	Ultra-chaos
Maximum value of λ_{max}	1.3×10^{-2}	8.7×10^{-2}
Minimum value of λ_{max}	8.5×10^{-5}	4.3×10^{-2}
Mean of λ_{max}	9.7×10^{-4}	6.9×10^{-2}
Standard deviation of λ_{max}	7.5×10^{-4}	1.0×10^{-2}

Lyapunov exponents λ_{max} of the trajectories are summarized in Table 5.2. Statistically speaking, the values of the maximum Lyapunov exponent λ_{max} of ultra-chaotic fluid particle motions in ABC flow are about two orders of magnitude larger than those of their normal-chaotic counterparts.

Let Θ denote the ratio of particles with ultra-chaotic trajectories to the total number of particles in the spatial domain $-\pi \le x, y, z \le +\pi$ of the ABC flow field specified by Eq. (5.14). Using the Monte-Carlo method to randomly choose 10000 starting fluid particles in the spatial domain $-\pi \le x, y, z < +\pi$, it is found that Θ increases as C enlarges, as shown in Table 5.3. For the special case of $A = 1$ and $B = 0.7$, a power-law relationship exists such that

$$\Theta \approx C^{0.4}, \qquad \text{when } C \le 0.1. \tag{5.18}$$

The foregoing results indicate that ultra-chaos can be widespread in ABC flow. For more details, please refer to Qin and Liao [57].

TABLE 5.3

Ratio Θ of particles with ultra-chaotic trajectories to the total number of particles in ABC flow (5.14) for $A = 1.0$, $B = 0.7$, and $0 \le C \le 0.43$

C	Θ
0.43	49%
0.20	47%
0.10	43%
10^{-2}	20%
10^{-3}	6%
10^{-4}	2%
0	0%

5.3 Possible Relationships of Ultra-chaos to Turbulence, Poincaré Section and Ergodicity

5.3.1 Ultra-chaos and Turbulence

To study the possible relationship between ultra-chaos and turbulence, let us consider the Navier–Stokes momentum equation with an external force and the continuity equation given by:

$$\frac{\partial \mathbf{u}}{\partial t} + (\mathbf{u} \cdot \nabla)\mathbf{u} = -\nabla p + \frac{1}{Re}\Delta \mathbf{u} + \mathbf{f}, \tag{5.19}$$

$$\nabla \cdot \mathbf{u} = 0, \tag{5.20}$$

subject to periodic boundary conditions at $x = \pm\pi$, $y = \pm\pi$, and $z = \pm\pi$, where $\mathbf{u}$ is the velocity vector, p is pressure, t is time, ∇ is the Hamilton operator, Δ is the Laplace operator, Re is the Reynolds number, and

$$\mathbf{f} = \frac{\mathbf{u}_{ABC}}{Re} \tag{5.21}$$

is the given external force per unit mass, in which $\mathbf{u}_{ABC}$ is the velocity of the ABC flow (5.14). As pointed out by Arnold [90], $\mathbf{u}_{ABC}$ is a steady-state solution of the NS and continuity equations (5.19) and (5.20). Besides, the Reynolds number $Re = 50$ corresponds to turbulent flow (when $C \neq 0$) [96], if small disturbances are added to the initial velocity field $\mathbf{u}_{ABC}$. So, in the remaining part of this subsection, we select $\mathbf{u}_{ABC}$ plus white noise of order of magnitude 10^{-3} as the initial guess and set the Reynolds number to $Re = 50$. The NS and continuity equations are solved numerically by first discretizing the spatial domain $[-\pi, +\pi)^3$ onto a uniform mesh with 128^3 points of a spatial Fourier expansion, where the maximum grid spacing is less than the minimum Kolmogorov scale [97], and then applying CNS with a time step of $\Delta t = 10^{-3}$. For more details, please refer to Qin and Liao [57].

The ABC flow is based on three parameters, A, B, and C. Without loss of generality, let us first consider the *unstable* ABC flow obtained for $A = 1, B = 0.7$ and $C = 0.43$, whose 49% of the fluid particles move along ultra-chaotic trajectories, according to Table 5.3. So, initially about half of the fluid particles of this flow move along ultra-chaotic trajectories. Transition from laminar to turbulent flow occurs at $t \approx 50.0 = T_{tran}$, where T_{tran} is the time at which transition occurs. At $t = 50$, using the Monte-Carlo method to randomly choose 10000 starting fluid particles in the spatial domain $-\pi \leq x, y, z < +\pi$, it is found that nearly *all* trajectories are ultra-chaotic in the resulting turbulent flow. This finding strongly suggests that a necessary precondition for turbulence governed by (5.19) and (5.20) is that almost all fluid particles follow ultra-chaotic trajectories.

Now let us consider *stable* ABC flow when $A = 1$, $B = 0.7$ and $C = 0$. In this case transition from laminar flow to turbulence *never* occurs, and *no* fluid particle exhibits ultra-chaotic motion. This further confirms the finding that turbulence is related to ultra-chaotic motions of fluid particles.

Next consider ABC flows for $A = 1$, $B = 0.7$, and varying C such that $0 < C \leq 0.43$. Again, all fluid particle trajectories are ultra-chaotic when the flow becomes fully turbulent. Besides, the transition time T_{tran} increases as C decreases. Thus, the smaller the number of ultra-chaotic fluid particles at the beginning (corresponding to an unstable ABC flow with smaller $C > 0$, for example as listed in Table 5.3), the longer the transition time T_{tran}. In other words, more time is needed for all fluid particles to become ultra-chaotic. Notably, the following linear relationship holds:

$$T_{tran} \approx -10 \log_{10}(C) + 40, \qquad \text{when } 0 < C \leq 0.1. \qquad (5.22)$$

According to (5.22), $T_{tran} \rightarrow +\infty$ as $C \rightarrow 0$, thus transition to turbulence cannot occur when $C = 0$. This concurs with the previous result for $C = 0$. These findings strongly suggest that nearly all fluid particles should move along ultra-chaotic trajectories when transition to turbulence occurs.

It should be emphasized that the findings in this subsection are purely based on the NS and continuity equations (5.19) and (5.20), in the presence of an external force (5.21). A question arises. Do all fluid particles actually move along ultra-chaotic trajectories in general cases of fully turbulent flow? This is an open question, which deserves further study. Note that ultra-chaos is a new concept, which might provide us with a new perspective on turbulence, especially the transition from laminar to turbulent flow that is one of the core issues in fluid mechanics.

5.3.2   Ultra-chaos and Poincaré Section

Figure 5.12 shows the Poincaré section (at $z = 0$) of ABC flow (5.14) for $A = 1$, $B = 0.7$, and $C = 0.43$. In the figure, $x', y' \in [-\pi, \pi]$ are renormalized values of x, y obtained when $z = 2n\pi$ (corresponding to $z' = 0$) for an arbitrary integer n, using the periodic condition in the x and y directions [91].

The Poincaré section consists of elliptic islands (or KAM tori, marked in blue) and a chaotic sea (marked in red). It is widely believed [98] that points in an elliptic island correspond to either quasi-periodic orbits or weakly chaotic orbits, but points located in a chaotic sea relate to strongly chaotic orbits.

It is very interesting that the Poincaré section (at $z = 0$) of the ABC flow (5.14) is rather similar to the classification of trajectories of the fluid particles starting at $z = 0$, as shown in Figure 5.12. Here, the normal-chaotic starting points (at $z = 0$, marked in blue) of the ABC flow correspond to elliptic islands (or KAM tori), and the ultra-chaotic starting points (marked in red) to the chaotic sea. Such a relationship holds for almost all fluid particles in the ABC flow (5.14). For more details, please refer to Qin and Liao [57].

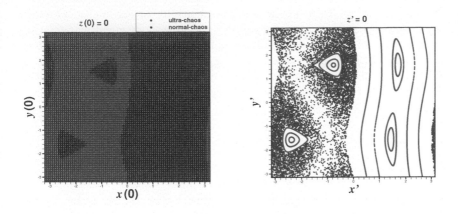

FIGURE 5.12

Comparison between the classification of trajectories and the Poincaré section of fluid particles (at $z = 0$) in ABC flow (5.14) for $A = 1$, $B = 0.7$, and $C = 0.43$. Left: classification of trajectories at $z = 0$; right: Poincaré section (at $z = 0$).

This finding confirms once again that ultra-chaos is indeed at a higher level of disorder than normal-chaos. Chaotic classification therefore provides a *new* explanation for the presence of elliptic islands (or KAM tori) and a chaotic sea in the Poincaré section.

Note that the Poincaré section is based on KAM theory. Generally speaking, KAM theory works for an *integrable* Hamiltonian system only. However, the classification of chaos into ultra-chaos and normal-chaos is generally valid for *all* dynamic systems, even if they are neither Hamiltonian nor integrable, and thus has a more general meaning!

Obviously, it is important to investigate other dynamic systems so as to confirm the general validity of the above-mentioned relationships between the classification of chaos and the Poincaré section.

5.3.3   Ultra-chaos and Ergodicity

Does normal-chaos correspond to an ergodic system? Does ultra-chaos correspond to a non-ergodic system? These are interesting questions.

It is well-known that time average is equal to phase average for an ergodic system [99, 100], i.e., a metrically transitive system [101]. Unfortunately, it is rather difficult in practice to prove definitively the metrical transitivity of a given dynamic system. Conversely, it is easy to classify a chaotic system according to the stability of its statistical results: the system exhibits normal-chaos if its statistics are stable, and ultra-chaos otherwise. From a practical viewpoint, it is much easier to check the stability of statistics of a given dynamic system by considering the tempo-spatial (or ensemble) average than to directly prove (or disprove) its metric transitivity or ergodicity. Our new classification of chaos provides us a simpler and more convenient way to study dynamic systems in practice, including their stability, ergodicity and so on.

It is still an open question whether or not normal-chaos corresponds to ergodicity and ultra-chaos to non-ergodicity. However, our new classification of chaos might provide a new way to study the ergodic property of dynamic systems.

5.4   Influence on Paradigm of Scientific Research

Modern science is based on the scientific method, which involves observations/questions, hypotheses, experimental tests, predictions, and so on. Ibn al Haytham (965 – 1040)[†], an early pioneer of the scientific method and the world's "first true scientist"[‡], suggested that a hypothesis must be supported by experiments which should be based on confirmable procedures. Galileo (1564 – 1642)[§], the father of modern science, who originally combined experiment and mathematics, set up standards of length and time so that experimental results measured on different days and in different laboratories could be compared in a *reproducible* way. Robert Boyle (1629 – 1691)[¶], a pioneer of the experimental method, believed that the foundations of knowledge should be constituted by experimental facts, made believable by their *reproducibility*[||], i.e., repeating the same experiment over and over again. From a historical perspective, reproducible experiments have indeed played a very important role in scientific advances. Without doubt, the reproducibility and/or replicability of experiments is a cornerstone of modern science.

For every experiment, artificial disturbance and/or environmental noise are *unavoidable* and out of the experimenter's control. Therefore, some small disturbances are present in every experiment. If a dynamic system is non-chaotic, then both its trajectory and statistics are stable, i.e., not sensitive to small disturbances. In such cases, both the trajectory and statistical results are reproducible. When a dynamic system experiences normal-chaos, its statistics remain stable even though its trajectory is unstable. In such cases, although the trajectories are not reproducible, the statistical results are (fortunately) reproducible. However, for an ultra-chaotic system, neither

---

[†] Please see https://en.wikipedia.org/wiki/Ibn_al-Haytham

[‡] Al-Khalili, Jim (4 January 2009). *The 'first true scientist'*. BBC News. Retrieved 2 June 2018. Please visit http://news.bbc.co.uk/2/hi/science/nature/7810846.stm

[§] Please see https://en.wikipedia.org/wiki/Galileo_Galilei

[¶] please see https://en.wikipedia.org/wiki/Robert_Boyle

[||] Please see https://en.wikipedia.org/wiki/Reproducibility

its trajectories nor its statistical results are stable: in this case, *neither* the trajectories *nor* statistical results are reproducible, and so it is *impossible* in practice to repeat a corresponding experiment or numerical simulation, even in a statistical sense! Notably, for an ultra-chaotic system, our paradigm of modern science, which is strongly dependent upon the reproducibility and/or replicability of physical/numerical experiments, becomes invalid: this is indeed a scientific catastrophe!

The reproducibility and/or replicability of experiments are based on such a principle that scientific laws are invariant across space and time. What happens if reproducibility or replicability of experiments does not exist, but is merely a mirage? As described in the first volume of Liu Cixin's famous Science Fiction trilogy "Remembrance of Earth's Past" [15] called "The Three-Body Problem", this indeed happens: "they repeated the ultra-high-energy collision experiments again and again using the same conditions, but every time the result was different", and as a result many scientists believe that "the laws of physics are *not* invariant across time and space". Note that, in Liu Cixin's science fiction novel, the loss of experimental reproducibility and/or replicability is caused by extraterrestrial living beings.

In fact, the so-called "crises of reproducibility and replicability" widely arise in numerous fields of science and engineering. Here, *reproducability* is defined at the ability to achieve the same results by repeating exactly the same experiments multiple times and *replicability* is defined as the ability to obtain the same results using different experimental facilities by different experimenters. According to a survey of 1,576 researchers in many fields by the journal *Nature*, more than half of the researchers failed to reproduce their own experiments, and it was concluded that there indeed exists a significant "crisis of reproducibility" [102]. An investigation into the replicability of published experiments in economics found that only 66% of the studies considered gave replicable results [103]. Although it is widely accepted that reproducibility and replicability in Computational Fluid Dynamics (CFD) serve as a minimum standard by which to judge scientific claims and discoveries [104], unfortunately, "completing a full replication study of previously published findings on bluff-body aerodynamics is harder than it looks, even when using good reproducible-research practices and sharing code and data openly" [105]. As reported by Peng [104], the replication ratio in computational science is only about 25%. Naturally, the non-reproducibility and non-replicability of more and more scientific research results has led to growing concern about the reliability of new scientific discoveries [106].

Practically speaking, there are lots of artificial reasons that could lead to irreproducible research results, such as poor experimental design, low statistical power, insufficient oversight/mentoring, selective reporting, unavailability of code & computational mesh, insufficient peer review, fraud, etc., as pointed by Baker *et al.* [102] and Mesnard *et al.* [105]. It should be emphasized that all of the foregoing reasons are *artificial*. But are there any *objective* reasons? Can we guarantee that all experiments that lack reproducibility and/or replicability do not belong to ultra-chaos? Note that statistical significance and p-values are based on reliable/replicable results of statistical analysis. However, for an ultra-chaotic system whose statistics are sensitive to tiny disturbances, it is practically impossible to replicate any new discoveries based on "statistically significant" findings. In theory, ultra-chaos might be an *objective* cause of non-reproducible research findings and the so-called "crisis of reproducibility".

Are all physical/numerical experiments under same controlled conditions reproducible and/or repeatable, at least in terms of statistics? How can we prove whether the answer is positive? Unfortunately, according to the author's best knowledge, there is no mathematical/physical proof available. Thus, it is merely a hypothesis that the statistics of all dynamic systems are stable to small disturbances. Thus, (5.2) is simply a *hypothesis* for a general dynamic system.

For an ultra-chaotic system, the modern scientific paradigm becomes invalid due to the loss of experimental reproducibility and/or replicability. Therefore, if ultra-chaotic systems really exist, our paradigm of modern science becomes incomplete. Such an incompleteness of scientific paradigm is akin to Gödel's incompleteness theorem in mathematical logic. Could such incompleteness shake the cornerstone of modern science? How should we understand and interpret the numerical/experimental results obtained from ultra-chaotic systems? How should we define "truth" in terms of ultra-chaos and what kind of "truth" could an ultra-chaos tell us? Certainly, there is still much work to be carried out in the future.

In this chapter, simulations of several dynamic systems have provided numerical evidence for the wide existence of ultra-chaos. However, it would be most welcome if mathematical/physical proofs about the existence of ultra-chaos could be given. Therefore, to conclude this chapter, the author lists an open question and two conjectures:

> **Open Question 5.1** *Statistics stability:* Can we mathematically prove (or disprove) that the statistical results of all dynamical systems are stable, i.e., not sensitive to small disturbances, so that $\langle \mathcal{P}+\delta' \rangle = \langle \mathcal{P} \rangle$ always holds when $\delta' \sim \mathcal{P}$ mostly? Here, $\langle \rangle$ denotes a statistical operator, $\mathcal{P}$ is a true physical solution, δ' denotes false noise caused by artificial and/or environmental disturbances, and $\mathcal{P} + \delta'$ denotes a solution trajectory as a mixture of $\mathcal{P}$ and δ' gained by conventional algorithms.

> **Conjecture 5.1** *Statistics instability:* There should exist certain dynamical systems, whose statistics are unstable, i.e., sensitive to small disturbances. In other words, $\langle \mathcal{P}+\delta' \rangle$ is not always equal to $\langle \mathcal{P} \rangle$, where $\langle \rangle$ denotes a statistical operator, $\mathcal{S}' = \mathcal{P} + \delta'$ is a numerical simulation of chaos obtained by conventional algorithms using single or double floating-point arithmetic, $\mathcal{P}$ denotes the true physical solution that can be accurately obtained by CNS, and δ' denotes false numerical noise, which all are mostly of the same order of magnitude, i.e., $\delta' \sim \mathcal{P} \sim \mathcal{S}'$.

Conjecture 5.2 *Incompleteness of modern scientific paradigm*: There should exist some dynamic systems with statistics instability, for which the current paradigm of modern science is invalid due to the loss of reproducibility and/or replicability of physical/numerical experiments.

6

Numerical Simulation of Turbulence: True or False?

Direct numerical simulation (DNS) [16, 17] is widely used as a numerical tool to investigate turbulence. To the best of the author's knowledge, all DNS results to date have been based on single (or double) precision floating-point arithmetic. As demonstrated in Section 4.2 and reported by Lin, Wang & Liao [44], spatiotemporal trajectories in turbulent flows are sensitive even to intrinsic thermal fluctuation (as the initial condition) and thus turbulent flows are essentially chaotic. Therefore, due to the butterfly-effect of chaos, a turbulent DNS result S' based on single (or double) precision floating-point arithmetic should quickly evolve into a mixture of the true physical solution $\mathcal{P}$ and false numerical noise δ', which are of the same order of magnitude, say $\delta' \sim \mathcal{P} \sim S'$. Thus, generally speaking, DNS results are often badly polluted by numerical noise δ'. Can a DNS result S' comprising such a mixture (i.e., $S' = \mathcal{P} + \delta'$) *always* agree satisfactorily with the true physical solution $\mathcal{P}$ from a statistical perspective, such that $\langle \mathcal{P} + \delta' \rangle = \langle \mathcal{P} \rangle$, where $\langle \rangle$ is a statistical operator? This open question is fundamental for turbulent flows.

In the frame of DNS [16, 17], the above open question is very hard to answer rigorously. This is mainly because there exists no way to separate the true physical solution $\mathcal{P}$ from the DNS result S' which is a mixture of the true physical solution $\mathcal{P}$ and false numerical noise δ' (such that $S' = \mathcal{P} + \delta'$). Thus, one hardly check whether or not the hypothesis

$$\langle \mathcal{P} + \delta' \rangle = \langle \mathcal{P} \rangle, \qquad \text{when } \delta' \sim \mathcal{P} \text{ mostly,} \qquad (6.1)$$

holds, where $\langle \rangle$ denotes a statistical operator. Conversely, as demonstrated by Lin, Wang & Liao [44], Clean Numerical Simulation (CNS) [38–40, 43, 54] can provide a convergent, reliable simulation S of a turbulent flow over a prescribed long time interval, during which the numerical noise δ' is many orders of magnitude lower than the true physical solution $\mathcal{P}$, i.e., $|\delta'| \ll |\mathcal{P}|$. In this case, the CNS trajectory S as a mixture of $\mathcal{P}$ and δ' is very close to the true physical solution $\mathcal{P}$, i.e., $S \approx \mathcal{P}$, and so the CNS result can provide sufficiently accurate statistics $\langle \mathcal{P} \rangle$. Comparing the statistical result $\langle \mathcal{P} \rangle$ obtained from the CNS benchmark solution with $\langle \mathcal{P} + \delta' \rangle$ obtained from conventional DNS [16, 17], one can directly check whether or not the hypothesis (6.1) holds.

DOI: 10.1201/9781003299622-6

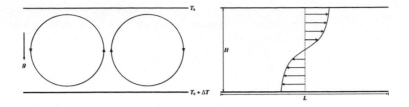

FIGURE 6.1
Two entirely different types of flow driven by two-dimensional turbulent Rayleigh-Bénard convection: vortical/roll-like flow (left), and zonal flow (right). Here, the fluid layer between two free surfaces (separated by a vertical distance H) entrains heat from the bottom boundary surface because of the constant temperature difference $\Delta T > 0$ between the upper and lower free surfaces. L denotes the horizontal length of the domain and g is the acceleration due to gravity.

In this chapter, comparisons are made between DNS and CNS results for two-dimensional turbulent Rayleigh-Bénard convection, as shown in Figure 6.1. It is found that numerical noise in the form of tiny artificial stochastic disturbances could indeed lead to large-scale deviations between the different simulations in terms not only of the spatiotemporal trajectories but also even of the statistics. Such numerical noise (in the form of artificial stochastic disturbances) can even lead to different types of flow. For example, shearing convection occurs in the DNS results with the corresponding flow field forming a kind of zonal flow pattern, whereas the CNS benchmark solution always sustains the non-shearing vortical/roll-like convection throughout the entire duration of the simulation. Thus, this provides us with rigorous evidence that numerical noise as a kind of small-scale artificial stochastic disturbance has quantitative and qualitative large-scale influences on sustained turbulence, exemplified by 2D turbulent RB convection.

6.1   Mathematical Model of Turbulent Rayleigh-Bénard Convection

As shown in Figure 6.1, let us consider two-dimensional turbulent Rayleigh-Bénard convection as a mathematical idealisation of sustained turbulence, subject to free-slip boundary conditions at the upper and lower surfaces and periodic boundary conditions at the lateral inflow/outflow boundaries in the horizontal direction. Setting the length scale H, velocity scale $\sqrt{g\alpha H\Delta T}$, and temperature scale ΔT as characteristic scales, the corresponding dimensionless Navier-Stokes (NS) momentum equations, combined with the Boussinesq approximation [78], may be written:

$$\frac{\partial}{\partial t}\nabla^2\psi + \frac{\partial(\psi, \nabla^2\psi)}{\partial(x, z)} - \frac{\partial\theta}{\partial x} - \sqrt{\frac{Pr}{Ra}}\nabla^4\psi = 0, \tag{6.2}$$

and

$$\frac{\partial\theta}{\partial t} + \frac{\partial(\psi, \theta)}{\partial(x, z)} - \frac{\partial\psi}{\partial x} - \frac{1}{\sqrt{PrRa}}\nabla^2\theta = 0, \tag{6.3}$$

subject to free-slip boundary conditions at the upper $(z = 1)$ and lower $(z = 0)$ surfaces:

$$\psi = \frac{\partial^2\psi}{\partial z^2} = \theta = 0 \tag{6.4}$$

and periodic boundary conditions at the lateral ends of the domain in the horizontal direction

$$\psi(0, z) = \psi(\Gamma, z), \quad \theta(0, z) = \theta(\Gamma, z), \tag{6.5}$$

where t denotes time, $x \in [0, \Gamma]$ and $z \in [0, 1]$ are horizontal and vertical (upwards positive) spatial coordinates, ψ is a stream-function defined according to its relationships to the horizontal and vertical velocity components as follows

$$u = -\frac{\partial\psi}{\partial z}, \qquad w = \frac{\partial\psi}{\partial x}, \tag{6.6}$$

θ is the temperature departure from a linear variation background*, $\Gamma = L/H$ is an aspect ratio, ∇^2 is the Laplace operator with the definition $\nabla^4 = \nabla^2\nabla^2$, and

$$\frac{\partial(a, b)}{\partial(x, z)} = \frac{\partial a}{\partial x}\frac{\partial b}{\partial z} - \frac{\partial b}{\partial x}\frac{\partial a}{\partial z} \tag{6.7}$$

is the Jacobi operator. The Rayleigh number Ra and Prandtl number Pr are defined by

$$Ra = \frac{g\alpha H^3\Delta T}{\nu\kappa}, \qquad Pr = \frac{\nu}{\kappa}, \tag{6.8}$$

respectively, in which α is the thermal expansion coefficient, $\nu = \mu/\rho$ is the fluid kinematic viscosity, and g is the acceleration due to gravitation.

Without loss of generality, let us consider the case when $\Gamma = 2\sqrt{2}$, $Pr = 6.8$ (corresponding to water at room temperature 20°C), and $Ra = 6.8 \times 10^8$ (corresponding to a turbulent state). Following Lin, Wang & Liao [44], the initial temperature and velocity fields correspond to intrinsic thermal fluctuations, which are randomly generated as Gaussian white noise with the temperature standard deviation $\sigma_T = 10^{-10}$ and the velocity standard deviation $\sigma_u = 10^{-9}$.

---

*The temperature is expressed as $T = \theta - z + 1$ in the case of $T_0 = 0$ on the upper plate.

6.2 CNS Algorithm in Physical Space

In 2017, Lin, Wang & Liao [44] combined clean numerical simulation (CNS) with the Fourier-Galerkin spectral method (in spectral space) to solve two-dimensional (2D) turbulent Rayleigh-Bénard convection (RBC) for free-slip boundary conditions. However, this kind of CNS algorithm in spectral space is rather time-consuming. In 2020, Hu & Liao [49] and Qin & Liao [50] proposed a more efficient CNS algorithm in physical space for spatiotemporal chaos, whose basic principles are briefly described below.

Substituting the coordinate transformations $\tilde{x} = \lambda x$ and $\tilde{z} = \mu z$ into Eqs. (6.2) and (6.3), where $\lambda = 2\pi/\Gamma$ and $\mu = \pi$, we have

$$
\frac{\partial}{\partial t}\left(\lambda^2 \psi_{xx} + \mu^2 \psi_{zz}\right) = \lambda\mu\psi_z\left(\lambda^2\psi_{xxx} + \mu^2\psi_{xzz}\right)
$$
$$
- \lambda\mu\psi_x\left(\lambda^2\psi_{xxz} + \mu^2\psi_{zzz}\right) + \lambda\theta_x
$$
$$
+ \sqrt{\frac{Pr}{Ra}}\left(\lambda^4\psi_{xxxx} + 2\lambda^2\mu^2\psi_{xxzz} + \mu^4\psi_{zzzz}\right), \quad (6.9)
$$

and

$$
\frac{\partial\theta}{\partial t} = \lambda\mu\left(\psi_z\theta_x - \psi_x\theta_z\right) + \lambda\psi_x + \frac{1}{\sqrt{PrRa}}\left(\lambda^2\theta_{xx} + \mu^2\theta_{zz}\right), \quad (6.10)
$$

for $t \geq 0$, $x \in [0, 2\pi]$ and $z \in [0, \pi]$, where the subscripts denote spatial derivatives and the overhead tildes are omitted (also for the remainder of this chapter).

To satisfy the free-slip boundary conditions at the lower ($z = 0$) and upper ($z = \pi$) surfaces using Fourier series, the computational domain is elongated from $z \in [0, \pi]$ to $z \in [0, 2\pi]$. We discretize the domain using $N_x \times N_z$ equidistant points, such that

$$
x_j = \frac{2\pi j}{N_x}, \qquad z_k = \frac{2\pi k}{N_z}, \qquad (6.11)
$$

where $j = 0, 1, 2, ..., N_x - 1$ and $k = 0, 1, 2, ..., N_z - 1$.

To reduce truncation errors in the temporal dimension to below a prescribed arbitrarily small level, Mth-order Taylor expansions are implemented, leading to the discretized stream function,

$$
\psi(x_j, z_k, t + \Delta t) \approx \sum_{m=0}^{M} \psi^{[m]}(x_j, z_k, t)(\Delta t)^m, \qquad (6.12)
$$

and the discretized temperature departure,

$$
\theta(x_j, z_k, t + \Delta t) \approx \sum_{m=0}^{M} \theta^{[m]}(x_j, z_k, t)(\Delta t)^m, \qquad (6.13)
$$

where Δt is the time-step, and

$$\psi^{[m]}(x_j, z_k, t) = \frac{1}{m!} \frac{\partial^m \psi(x_j, z_k, t)}{\partial t^m}, \tag{6.14}$$

$$\theta^{[m]}(x_j, z_k, t) = \frac{1}{m!} \frac{\partial^m \theta(x_j, z_k, t)}{\partial t^m}. \tag{6.15}$$

Differentiating both sides of Eqs. (6.9) and (6.10) $(m-1)$ times with respect to t and then dividing by $m!$ gives the following discretized governing equations for $\psi^{[m]}$ and $\theta^{[m]}$:

$$\lambda^2 \psi_{xx}^{[m]} + \mu^2 \psi_{zz}^{[m]} = \frac{1}{m} \left\{ \sqrt{\frac{Pr}{Ra}} \left[2\lambda^2 \mu^2 \psi_{xxzz}^{[m-1]} + \lambda^4 \psi_{xxxx}^{[m-1]} + \mu^4 \psi_{zzzz}^{[m-1]} \right] \right.$$

$$+ \sum_{r=0}^{m-1} \lambda \mu \psi_z^{[r]} \left[\lambda^2 \psi_{xxx}^{[m-1-r]} + \mu^2 \psi_{xzz}^{[m-1-r]} \right]$$

$$\left. - \sum_{r=0}^{m-1} \lambda \mu \psi_x^{[r]} \left[\lambda^2 \psi_{xxz}^{[m-1-r]} + \mu^2 \psi_{zzz}^{[m-1-r]} \right] + \lambda \theta_x^{[m-1]} \right\}, \tag{6.16}$$

and

$$\theta^{[m]} = \frac{1}{m} \left\{ \frac{1}{\sqrt{PrRa}} \left[\lambda^2 \theta_{xx}^{[m-1]} + \mu^2 \theta_{zz}^{[m-1]} \right] \right.$$

$$\left. + \lambda \mu \sum_{r=0}^{m-1} \psi_z^{[r]} \theta_x^{[m-1-r]} - \lambda \mu \sum_{r=0}^{m-1} \psi_x^{[r]} \theta_z^{[m-1-r]} + \lambda \psi_x^{[m-1]} \right\}, \tag{6.17}$$

where $m \geq 1$.

Note that Eqs. (6.16) and (6.17) contain spatial partial derivatives (denoted by x and z subscripts). In order to determine high precision approximations of these spatial partial derivative terms from known $\psi^{[r]}$ and $\theta^{[r]}$ at the discrete points $(x_j, z_k) \in \Omega$, we use the spatial Fourier series

$$\psi^{[r]}(x, z, t) \approx \sum_{n_x = -\frac{N_x}{2}+1}^{\frac{N_x}{2}-1} \sum_{n_z = -\frac{N_z}{2}+1}^{\frac{N_z}{2}-1} \Psi_{n_x, n_z}^{[r]}(t)\, e^{in_x x}\, e^{in_z z}, \tag{6.18}$$

and

$$\theta^{[r]}(x, z, t) \approx \sum_{n_x = -\frac{N_x}{2}+1}^{\frac{N_x}{2}-1} \sum_{n_z = -\frac{N_z}{2}+1}^{\frac{N_z}{2}-1} \Theta_{n_x, n_z}^{[r]}(t)\, e^{in_x x}\, e^{in_z z}, \tag{6.19}$$

where $i = \sqrt{-1}$ and the coefficients

$$\Psi_{n_x, n_z}^{[r]}(t) = \frac{1}{N_x N_z} \sum_{j=0}^{N_x-1} \sum_{k=0}^{N_z-1} \psi^{[r]}(x_j, z_k, t)\, e^{-in_x x_j}\, e^{-in_z z_k}, \tag{6.20}$$

and

$$\Theta_{n_x,n_z}^{[r]}(t) = \frac{1}{N_x N_z} \sum_{j=0}^{N_x-1} \sum_{k=0}^{N_z-1} \theta^{[r]}(x_j, z_k, t) \, e^{-i n_x x_j} \, e^{-i n_z z_k}, \qquad (6.21)$$

are given by the known $\psi^{[r]}(x_j, z_k, t)$ and $\theta^{[r]}(x_j, z_k, t)$ at the discrete points $(x_j, z_k) \in \Omega$. Accurate approximations of the spatial partial derivative terms in Eqs. (6.16) and (6.17) are then determined using Eqs. (6.18) and (6.19). Computational efficiency is enhanced by using the fast Fourier transform (FFT) algorithm and parallelized algorithms.

For a given value of $m \geq 1$, the right-hand sides of Eqs. (6.16) and (6.17) are known. Thus, $\theta^{[m]}$ at all discretized points can be directly obtained from Eq. (6.17). But $\psi^{[m]}$ is governed by the Poisson equation (6.16), which leads to a set of linear algebraic equations in terms of $\Psi_{n_x,n_z}^{[m]}$ by substituting (6.18) into (6.16). Thereafter, the discretized stream-function $\psi^{[m]}$ is given by (6.18) using the known coefficients $\Psi_{n_x,n_z}^{[m]}$. Similarly, it is then possible to determine $\theta^{[m+1]}$ and $\psi^{[m+1]}$, and so on.

As reported by Qin & Liao [55], the computational efficiency of the above-mentioned CNS algorithm in physical space is about three orders of magnitude higher than that of the CNS algorithm in spectral space [44]. For further information, please refer to Qin & Liao [55].

Obviously, provided the order M of the Taylor expansions (6.12) and (6.13) is sufficiently large, the temporal truncation error can be reduced to below a prescribed *arbitrarily* tiny level. Moreover, if the mode numbers N_x and N_z are also sufficiently large, the spatial truncation errors in the Fourier expressions (6.18) and (6.19) and spatial derivative terms in (6.16) and (6.17) can be reduced to below a required *arbitrarily* tiny level. However, this is *not* entirely sufficient, because round-off error is inevitable owing to the finite number of significant digits used to represent floating-point data. Therefore, *all* floating-point data must be expressed in multiple precision (MP) arithmetic using a sufficiently large number N_s of significant digits so as to reduce the round-off error to below a required *arbitrarily* tiny level. In this way, both the spatiotemporal truncation error and round-off error can simultaneously be reduced to below a required *arbitrarily* tiny level. Hence, the background numerical noise, determined as the maximum of the truncation error and round-off error, can be reduced to below a required *arbitrarily* tiny level. This approach is different from all other conventional numerical algorithms including DNS. In practice, CNS results are invariably more accurate than those given by DNS because CNS adopts multiple precision whereas DNS mostly uses double precision floating-point arithmetic[†].

[†]Please see https://en.wikipedia.org/wiki/Floating-point_arithmetic

6.3   Comparisons between CNS and DNS Trajectories

Qin & Liao [55] numerically solved Eqs. (6.2) and (6.3) by DNS and CNS, separately. Their DNS algorithm is based on a Runge-Kutta method in time and Fourier series in space, with all computations undertaken using double precision floating-point arithmetic (the simulation is called RKwD herein). In this DNS scheme, the numerical noise δ' increases quickly to the same order of magnitude as the true physical solution $\mathcal{P}$, say, $\delta' \sim \mathcal{P}$. For the CNS algorithm, the numerical noise δ' is much lower than the true physical solution $\mathcal{P}$ over the long time interval $t \in [0, T_c]$, during which $|\delta'| \ll |\mathcal{P}|$, and thus is negligible compared to the true physical solution $\mathcal{P}$. Consequently, $\langle S \rangle \approx \langle \mathcal{P} \rangle$ was found to hold for the CNS trajectory $S = \mathcal{P} + \delta'$, say, $\langle S \rangle$ given by CNS is a good approximation of $\langle \mathcal{P} \rangle$, where $\langle \rangle$ denotes a statistical operator.

First, we apply CNS in order to greatly decrease the background numerical noise to below such a tiny level that the exponentially increasing numerical noise is much smaller than the true physical solution $\mathcal{P}$ of the turbulent RBC over an interval of time that is sufficiently long for statistical analysis of the results to be accurate. Briefly speaking, in order to decrease the spatial truncation-error to a low enough level, the spatial domain is discretized by a uniform mesh $N_x \times N_z = 1024 \times 1024$, and a Fourier spectral solver with 3/2 rule for dealiasing [97] used. The spatial resolution is sufficiently high to represent properly the turbulent RBC under consideration: the horizontal (maximum) grid spacing $\Delta_x = L/N_x = 0.00276$ is smaller than the minimum Kolmogorov scale [97], as shown later. Besides, in order to decrease the temporal truncation-error to below a prescribed small level, a 45th-order (i.e., $M = 45$) Taylor expansion is used with time step $\Delta t = 10^{-3}$. In order to decrease the round-off error to below a sufficiently small level, all computations are undertaken using multiple precision (MP) floating-point arithmetic with 70 significant digits (i.e., $N_s = 70$). Another CNS result is then obtained using the Fourier spectral method on the same uniform mesh $N_x \times N_z = 1024 \times 1024$ but even smaller background numerical noise by means of a higher order (i.e., $M = 47$) Taylor expansion with the same time step ($\Delta t = 10^{-3}$) and using higher multiple precision floating-point arithmetic with more significant digits (i.e., $N_s = 72$). Comparing these two CNS results, it is found that no distinct differences are evident over the time interval $0 \leq t \leq 500$, which is long enough for statistical analysis. This verifies convergence and reliability of the CNS result during $t \in [0, 500]$ obtained with $M = 45$, $\Delta t = 10^{-3}$, and $N_s = 70$, and so this is taken as the "clean" benchmark solution hereafter.

Secondly, using the *same* initial/boundary conditions and the *same* physical parameters, Eqs. (6.2) and (6.3) are now solved numerically using a conventional algorithm based on the fourth-order Runge-Kutta method in time (with time-step $\Delta t = 10^{-4}$) and double Fourier series in space with double precision floating-point arithmetic (called RKwD). In this case, the numerical noise δ' increases quickly to the same level as the true physical solution $\mathcal{P}$ and so is not negligible. By comparing these RKwD simulations with the CNS benchmark solution, Qin & Liao [55] studied the influence of numerical noise as tiny artificial stochastic disturbances on the 2D turbulent RBC in detail. Their findings are outlined below.

Figures 6.2 and 6.3 compare the numerical simulation S' given by RKwD with the "clean" benchmark solution S given by CNS. Both simulations commence from exactly the same initial conditions which represent micro-level thermal fluctuations. For both the CNS and RKwD simulations, the thermal fluctuations in velocity and temperature as tiny initial disturbances evolve progressively from micro-level to macro-level until $t \approx 25$ when transition from laminar flow to turbulence occurs. Strong mixing then occurs during $t \in [25, 36]$ and a typical vortical/roll-like convection pattern appears at $t \approx 50$, as shown in Figure 6.2. Thereafter, the RKwD simulation S' increasingly deviates from the CNS benchmark solution S with distinct differences becoming obvious in their thermo-fluid structures, indicating that the numerical noise (in the form of artificial stochastic disturbances) could indeed lead to *large-scale* differences in velocity and temperature at a macroscopic level. Figure 6.2 shows such an example at times $t = 100$ and $t = 185$. This qualitatively confirms the conclusions of Lin et al. [44]. Even so, the flow fields obtained for the two simulations retain the vortical/roll-like turbulent convection patterns until $t \approx 188$ when the patterns diverge completely. In the RKwD simulation S', shearing convection then occurs and a zonal flow field emerges, as shown in Figures 6.2 and 6.3. Conversely, the CNS benchmark solution S sustains the non-shearing vortical/roll-like convection pattern throughout the remainder of the simulation process. In short, the RKwD simulation S' and CNS benchmark solution S exhibit completely different *types* of turbulent convection after $t > 188$. It should be emphasized that this kind of *qualitative* deviation in the large-scale flow structure is triggered merely by numerical noise (originating from identical artificial stochastic disturbances). The foregoing results strongly suggest that numerical noise (arising from artificial stochastic disturbances) exerts both quantitative and qualitative *large-scale* influences on *sustained* turbulence in our considered case of two-dimensional RB convection. For details, please refer to Qin & Liao [55].

6.4 Influence of Small Disturbances on Statistics

What about the influence of numerical noise (in the form of artificial stochastic disturbances) on the statistics of the 2D Rayleigh-Bénard convection results?

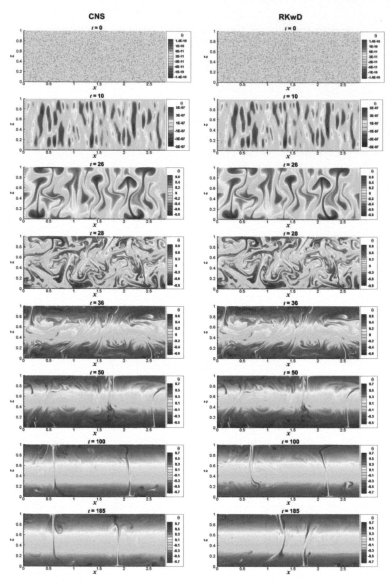

FIGURE 6.2

Comparison of the temperature fields θ (as departures from the linear variation background) during $0 \leq t \leq 185$ for $Pr = 6.8$, $Ra = 6.8 \times 10^8$, and $L/H = 2\sqrt{2}$. Left: benchmark solution given by CNS; right: DNS simulation obtained using Runge-Kutta method ($\Delta t = 10^{-4}$) in double precision floating-point arithmetic.

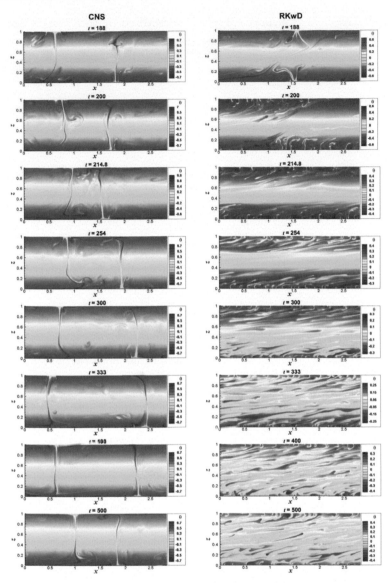

FIGURE 6.3

Comparison of the temperature fields θ (as departures from the linear variation background) during $188 \leq t \leq 500$ for $Pr = 6.8$, $Ra = 6.8 \times 10^8$, and $L/H = 2\sqrt{2}$. Left: benchmark solution given by CNS; right: DNS simulation obtained using Runge-Kutta method ($\Delta t = 10^{-4}$) in double precision floating-point arithmetic.

Convective heat transport is typically described by the Nusselt number

$$Nu(t) = 1 - \left.\frac{\partial\langle\theta(x,z,t)\rangle_x}{\partial z}\right|_{z=1}, \tag{6.22}$$

where the statistical operator

$$\langle f \rangle_x = \frac{1}{\Gamma}\int_0^\Gamma f\,dx \tag{6.23}$$

denotes the horizontal spatial average. The global convection strength is measured by the Reynolds number,

$$Re(t) = \sqrt{\frac{Ra}{Pr}}\,U_{rms}, \tag{6.24}$$

where $U_{rms} = \sqrt{\langle u^2 + w^2\rangle_A}$ is the root-mean-square (rms) velocity [107, 108], and the statistical operator

$$\langle f \rangle_A = \frac{1}{\Gamma}\int_0^\Gamma\int_0^1 f\,dx\,dz \tag{6.25}$$

denotes the spatial average. Figure 6.4 shows that both the $Nu(t)$ and $Re(t)$ time series given by the RKwD simulation S' agree well with the corresponding CNS results until $t \approx 188$ when the RKwD results begin to diminish greatly by comparison to the CNS results. The huge difference is not only *quantitative* but also *qualitative*, and is entirely due to the appearance of the zonal flow at $t \approx 188$ in the RKwD flow field, triggered by the exponential increase of false numerical noise in the RKwD simulation S' (arising from the artificial stochastic disturbances).

Let us also consider two important small-scale properties of the fluid flow: namely, the kinetic energy dissipation rate $\langle\varepsilon_V\rangle_A$ and the thermal dissipation rate $\langle\varepsilon_T\rangle_A$, defined by

$$\varepsilon_V(x,z,t) = \frac{1}{2}\sqrt{\frac{Pr}{Ra}}\sum_{i,j}\left[\partial_i u_j(x,z,t) + \partial_j u_i(x,z,t)\right]^2 \tag{6.26}$$

and

$$\varepsilon_T(x,z,t) = \frac{1}{\sqrt{Pr\,Ra}}\left|\nabla\left[\theta(x,z,t) - z\right]\right|^2, \tag{6.27}$$

in which $i, j = 1, 2$, $u_1 = u$, $u_2 = w$, $\partial_1 = \partial/\partial x$, $\partial_2 = \partial/\partial z$, ∇ is the Hamilton operator, and $\langle\ \rangle_A$ denotes a statistical operator of the spatial average defined by (6.25). As shown in Figure 6.5, for the RKwD simulation S', both $\langle\varepsilon_V\rangle_A$ and $\langle\varepsilon_T\rangle_A$ fall sharply when $t \approx 188$ at the onset of shearing convection (i.e., zonal flow), triggered by the growth of numerical noise (from artificial

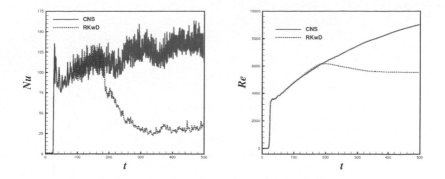

FIGURE 6.4

Time histories of (left) instantaneous Nusselt number Nu and (right) Reynolds number Re obtained for $Pr = 6.8$, $Ra = 6.8 \times 10^8$, and $L/H = 2\sqrt{2}$. Blue solid line: CNS benchmark solution; red dashed line: DNS simulation given by Runge-Kutta method ($\Delta t = 10^{-4}$) using double precision floating-point arithmetic.

stochastic disturbances) in the RKwD simulation. For more details, please refer to Qin & Liao [55].

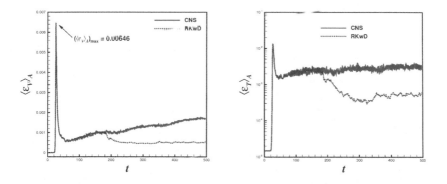

FIGURE 6.5

Time histories of (left) spatially averaged kinetic energy dissipation rate $\left\langle \varepsilon_V \right\rangle_A$ and (right) thermal dissipation rate $\left\langle \varepsilon_T \right\rangle_A$ obtained for $Pr = 6.8$, $Ra = 6.8 \times 10^8$, and $L/H = 2\sqrt{2}$, where $\left\langle f \right\rangle_A$ is defined by (6.25). Blue solid line: CNS benchmark solution; red dashed line: DNS simulation given by Runge-Kutta method ($\Delta t = 10^{-4}$) using double precision floating-point arithmetic.

Figure 6.5 (left) shows that the maximum kinetic energy dissipation rate $(\langle \varepsilon_V \rangle_A)_{max} = 0.00646$ occurs at $t = 26.9$ just as the transition from the laminar flow to turbulence starts to occur, corresponding to the minimum Kolmogorov scale

$$\left(\langle \eta \rangle_A \right)_{min} \approx \left(\frac{Pr}{Ra} \right)^{3/8} \left[\left(\langle \varepsilon_V \rangle_A \right)_{max} \right]^{-1/4} = 0.00353.$$

Thus, the criterion that the maximum grid spacing

$$\Delta_x = \frac{\Gamma}{N_x} = 0.00276 < 0.8 \left(\langle \eta \rangle_A \right)_{min} = 0.00282$$

is satisfied. Hence, the spatial resolution used herein is physically fine enough for accurate simulation of our case of two-dimensional Rayleigh-Bénard convection.

As reported by McMullen *et al.* [109], stochastic thermal fluctuation might influence the *small-scale* properties of *freely* decaying turbulence. As we have previously seen, numerical noise as a kind of micro-level artificial stochastic disturbance could lead to *large-scale* deviations not only in spatiotemporal trajectories but also in the statistics of two-dimensional *sustained* turbulent Rayleigh-Bénard convection. Through numerical rigour, Qin & Liao [55] have provided us with *convincing* evidence that numerical noise derived from tiny artificial stochastic disturbances exerts both quantitative and qualitative *large-scale* impacts on *sustained* turbulence. It is recommended that more investigations like the foregoing be carried out in the future for a whole range of turbulent flows. For more details, please refer to Qin & Liao [55].

6.5   DNS of Turbulence: True or False?

In Sections 6.3 and 6.4, the CNS result is used as the benchmark solution for comparisons against DNS results obtained using double precision floating-point arithmetic. It is found that distinct deviations in the two sets of thermo-fluid statistics start to occur at $t \approx 188$ as the DNS flow field alters to shearing convection but the CNS flow field retains its vortical/roll-like (non-shearing) convection over the whole time interval $t \in [0, 500]$.

Due to the butterfly-effect, a numerical simulation of turbulent flow comprises a mixture of the true physical solution $\mathcal{P}$ and false numerical noise δ'. For the above DNS trajectory $\mathcal{S}'$ of RB convection obtained using double precision floating-point arithmetic, the false numerical noise δ' exponentially increases, and reaches the *same* order of magnitude as the true physical solution $\mathcal{P}$ at $t \approx 50$, i.e., $\delta' \sim \mathcal{P}$, as shown in Figure 6.6. Thereafter, the DNS spatiotemporal trajectories (i.e., the mixture $\mathcal{S}' = \mathcal{P} + \delta'$) of the temperature and velocity fields are badly polluted by numerical noise δ' and thus *distinctly*

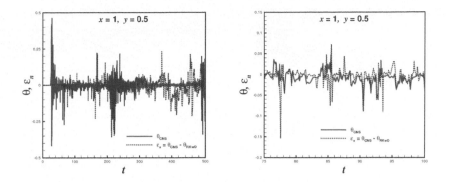

FIGURE 6.6

Comparison between CNS benchmark solution θ_{CNS} and numerical noise $\varepsilon_n = \theta_{CNS} - \theta_{RKwD}$ at the probe point location $x = 1$ and $y = 1/2$, where θ_{RKwD} denotes the RKwD simulation. Blue solid line: CNS benchmark solution; red dashed line: numerical noise. Left: $1 \leq t \leq 500$; right: $75 \leq t \leq 100$.

deviate from the true physical solution $\mathcal{P}$. Conversely, the background numerical noise in the CNS trajectory $\mathcal{S}$ is so small that the exponentially increasing false numerical noise δ' remains much smaller than the true physical solution $\mathcal{P}$ during the *whole* time interval $t \in [0, 500]$, i.e., $|\delta'| \ll |\mathcal{P}|$, so that the CNS trajectory $\mathcal{S}$ throughout $t \in [0, 500]$ can be used as an accurate approximation of the true physical solution $\mathcal{P}$, and so $\mathcal{S} = \mathcal{P} + \delta' \approx \mathcal{P}$ *does* hold for the CNS trajectory $\mathcal{S}$ as a mixture of $\mathcal{P} + \delta'$. It is a striking phenomenon that the statistics of DNS and CNS results agree qualitatively until $t \approx 188$ when shearing convection occurs in the DNS (but not CNS) flow field, as shown in Figures 6.4 and 6.5.

Why does this phenomenon occur? Qin and Liao [55] provide the following explanation. The 2D turbulent RB convection under consideration has two possible states, vortical/roll-like flow and shearing flow, which can be regarded as the two minima of a double-well potential. Once the CNS benchmark solution encounters one of these minima (e.g. roll-like flow), the solution remains there forever. This is because the corresponding false numerical noise δ' is much less than the true physical solution $\mathcal{P}$ and thus can *not* trigger a further transition to another minimum. Conversely, the DNS simulation can trigger a transition from one state to the other because its false numerical noise δ' arising from artificial stochastic disturbance can reach the same order of magnitude as the true physical solution $\mathcal{P}$, as shown in Figure 6.6, in which case the DNS simulation might depart very far away from its true physical solution $\mathcal{P}$ and can transition to the other minimum. Such transitions of the DNS predictions from one state to another should occur randomly for

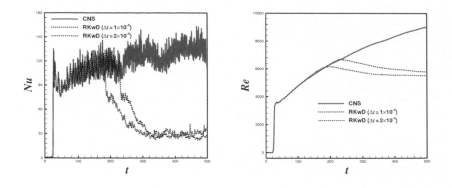

FIGURE 6.7

Evolution of (left) Nusselt number Nu and (right) Reynolds number Re for $Pr = 6.8$, $Ra = 6.8 \times 10^8$, and $L/H = 2\sqrt{2}$. Blue solid line: CNS benchmark solution; red dashed line: DNS simulation (with $\Delta t = 10^{-4}$); black dashed line: DNS simulation (with $\Delta t = 2 \times 10^{-4}$).

different levels of numerical noise. This is indeed true: as shown in Figure 6.7, transition from vortical/roll-like flow to zonal flow in the DNS results occurs at $t \approx 188$ and $t \approx 230$ for $\Delta t = 10^{-4}$ and $\Delta t = 2 \times 10^{-4}$, respectively. This evidence backs the above explanation, and provides us with useful information by which to understand better the origin of transition to turbulence in fluid flows.

Thus, generally speaking, if a turbulent flow has multiple large-scale states, then small disturbances, which are either natural (such as environmental perturbations) or artificial (such as background numerical noise), could lead to random transitions between different states. In the case we have just considered it is *impossible* to make a correct prediction of flow states by means of DNS in single (or double) precision floating-point arithmetic. However, if a turbulent flow has a *unique* state, background numerical noise should not exert a large-scale influence on the flow in statistics. This is of course good news for researchers in the field of CFD.

It is an open question as to whether or not micro-scale thermal fluctuations (analogous to tiny artificial numerical noises) might evolve to have a large-scale impact on certain turbulent flows with multiple states. It is strongly recommended that studies be carried out on the large-scale influence of thermal fluctuations on turbulent flows represented by the Landau-Lifshitz-Navier-Stokes (LLNS) equations [110]. It is also worthwhile investigating which set of equations gives the more precise description of turbulent flows (especially those with large-scale multiple states), either the *deterministic* NS equations or the *stochastic* LLNS equations.

In summary, numerical noise as weak, small-scale stochastic perturbations increases exponentially to a macro-level in numerical simulations, and can exert great influence on the long-term macroscopic statistics of turbulent flows. Therefore, we must pay very careful attention to the reliability of DNS of turbulence, especially when applied to cases involving large-scale multiple states.

Note that the critical predictable time T_c is an important concept in the frame of CNS, since a CNS result is *only* convergent in a limited time interval $t \in [0, T_c]$ and one had to spend a lot to enlarge T_c for a chaotic system. Conversely, direct numerical simulation (DNS) [16, 17] has no such a concept at all, since one can gain a simulation S' by DNS in a time interval as long as one would like: this however is based on such a hypothesis that the corresponding statistics are stable, say, not sensitive to small disturbances including numerical noises, i.e.,

$$\langle \mathcal{P} + \delta' \rangle = \langle \mathcal{P} \rangle, \quad \text{when } \delta' \sim \mathcal{P} \text{ mostly,}$$

where $\mathcal{P}$ is the true physical solution and δ' is the numerical noise, which are mostly of the same order of magnitude, say, $\delta' \sim \mathcal{P} \sim S'$. Unfortunately, this hypothesis does not hold for ultra-chaotic systems, which widely exist in nonlinear dynamics, as illustrated in Chapter 5 of this book. Thus, the use of badly polluted DNS results has a precondition: the statistics stability, corresponding to normal-chaotic systems. So, ultra-chaos as a new concept also reveals the precondition and limitation of DNS [16, 17].

6.6 Modified Fourth Clay Millennium Problem

In 2000 the Clay Mathematics Institute announced seven millennium problems[‡]. The fourth Clay millennium problem[§] concerns the existence and smoothness of the Navier–Stokes equation.

Environmental disturbances are *not* considered in the fourth Clay millennium problem. However, as illustrated in Section 4.2, spatiotemporal trajectories of Rayleigh-Bénard turbulent convection governed by the Navier–Stokes equation exhibit sensitivity dependence on the initial condition (SDIC), which strongly implies that turbulent flows belong to chaos in general. This chapter demonstrates that a tiny level of background (artificial) numerical noise can lead to significant deviations in numerical simulations of Rayleigh-Bénard turbulent convection governed by the Navier–Stokes and continuity equations

---

[‡]See the website http://www.claymath.org/millennium-problems/.

[§]Official description: http://www.claymath.org/sites/default/files/navierstokes.pdf.

not only in statistics but also even in the type of macroscopic flow. It should of course be emphasized that small environmental disturbances are unavoidable in practice. Although such small disturbances can have macroscopic effects on viscous flows, they are unfortunately neglected in the fourth Clay millennium problem concerning the Navier–Stokes equation.

Background numerical noise from artificial sources is unavoidable when the Navier–Stokes and continuity equations are solved numerically. In practice, it is often not very important whether or not numerical spatiotemporal trajectories of turbulent flows governed by the Navier–Stokes and continuity equations are sensitive to numerical noise. However, it is very important whether or not the *statistical* results of such simulations are sensitive to numerical noise. So, the author strongly recommends that the following open question be considered as a *modified fourth Clay millennium problem*:

> **Open question 6.1** *Existence, smoothness, and statistics stability of the Navier–Stokes and continuity equations*: Can we prove the existence and smoothness of the solution of the Navier–Stokes and continuity equations for physically sound boundary and initial conditions, where the statistics of the solution are stable (or unstable), in other words insensitive (or sensitive) to small disturbances?

It is well-known that the fourth Clay millennium problem remains an open question. In mathematics, the modified fourth Clay millennium problem is likely to be more difficult than its original counterpart. In practice, however, statistics stability is more important than smoothness, especially for researchers working in the field of computational fluid dynamics (CFD); otherwise, it is necessary to consider carefully the influence of unavoidable artificial numerical noise on the statistics of the numerical simulations! In other words, the precondition for the application of conventional Direct Numerical Simulation (DNS) on turbulent flow is that the flow is a normal-chaos. Therefore, the modified fourth Clay millennium problem has important physical meaning.

To capture the influence of thermal fluctuations on fluid flows at a macroscopic hydrodynamic scale, a stochastic form of the NS equation was proposed by Landau & Lifshitz [110], namely the Landau-Lifshitz-Navier-Stokes (LLNS) equation [111–114]:

$$\frac{\partial \mathbf{U}}{\partial t} + \nabla \cdot \mathbf{F}_a = \nabla \cdot \mathbf{F}_b + \nabla \cdot \mathbf{F}_s, \qquad (6.28)$$

subject to properly specified boundary and initial conditions, the latter given by

$$\mathbf{U}\big|_{t=0} = \mathbf{U}_0(\mathbf{r}), \qquad \mathbf{r} \in \Omega, \qquad (6.29)$$

where $\mathbf{U}(\mathbf{r}, t) = (\rho, \rho\mathbf{u}, E)^{\mathrm{T}}$ is a vector of variables under consideration, $\mathbf{r} \in \Omega$ is the vector of spatial position, t denotes time, $\mathbf{U}_0(\mathbf{r})$ is a specified initial vector of the variables under consideration, ρ is mass density, $\mathbf{u}$ is the velocity vector, E is total energy per unit mass, ∇ is the Hamiltonian/gradient operator,

$$\mathbf{F}_a = \begin{pmatrix} \rho\,\mathbf{u} \\ \rho\,\mathbf{u} \otimes \mathbf{u} + P\mathbf{I} \\ (E+P)\mathbf{u} \end{pmatrix}, \qquad \mathbf{F}_b = \begin{pmatrix} 0 \\ \tau \\ \mathbf{u}\cdot\tau + \mathbf{q} \end{pmatrix} \qquad (6.30)$$

are advection/hyperbolic and diffusion fluxes, $\mathbf{u} \otimes \mathbf{u}$ denotes the tensor product of the velocity vector $\mathbf{u}$, P is pressure, $\mathbf{I}$ is the unit tensor, τ is the tensor of viscous stress, $\mathbf{q}$ is the vector of heat diffusion flux, and the stochastic flux

$$\mathbf{F}_s = \begin{pmatrix} 0 \\ \mathbf{S} \\ \mathbf{u}\cdot\mathbf{S} + \mathbf{Q} \end{pmatrix} \qquad (6.31)$$

is composed of the stochastic stress tensor $\mathbf{S}$ and the stochastic heat flux vector $\mathbf{Q}$, whose magnitudes are determined by the fluctuation-dissipation theorem

$$\langle \mathbf{S}_{i,j}(\mathbf{r},t) \rangle = \langle \mathbf{Q}_k(\mathbf{r},t) \rangle = \langle \mathbf{S}_{i,j}(\mathbf{r},t)\mathbf{Q}_k(\mathbf{r}',t') \rangle = 0, \qquad (6.32)$$

$$\langle \mathbf{S}_{i,j}(\mathbf{r},t)\mathbf{S}_{m,n}(\mathbf{r}',t') \rangle$$
$$= 2\mu k_B T\,\delta(\mathbf{r}-\mathbf{r}')\delta(t-t')\left(\delta_{i,m}\delta_{j,n} + \delta_{i,n}\delta_{j,m} - \frac{2}{3}\delta_{i,j}\delta_{m,n}\right), \qquad (6.33)$$

and

$$\langle \mathbf{Q}_i(\mathbf{r},t)\mathbf{Q}_j(\mathbf{r}',t') \rangle = 2\kappa k_B T^2\,\delta(\mathbf{r}-\mathbf{r}')\delta(t-t')\delta_{i,j}, \qquad (6.34)$$

where i, j, k, m, n are subscripts of tensor/vector components, $\langle\,\rangle$ is the averaging operator, μ is the shear viscosity, k_B is Boltzmann's constant, T is fluid temperature, κ is the coefficient of thermal diffusivity, $\delta(x)$ is the Dirac delta function, and $\delta_{i,j}$ is the Kronecker delta function.

Note that, when stochastic fluxes are neglected, i.e., $\mathbf{F}_s = 0$, the LLNS equation (6.28) reduces to the NS equation. In other words, the NS equation in the fourth Clay millennium problem is simply a special case of the LLNS equation (6.28). Therefore, the LLNS equation provides us with an ideal mathematical model by which to investigate the influence of thermal fluctuations on turbulent flows.

The fourth Clay millennium problem concerns the existence and smoothness of the Navier–Stokes equation under proper boundary and initial conditions. In mathematics, solution existence is also very important for the LLNS equation (6.28). However, from a practical standpoint, it is more important to know the *stability* of the statistical results of the LLNS equation (6.28) under small disturbances such as thermal fluctuations. Thus, the author recommends a second open question:

Open Question 6.2 *Existence and statistics stability of the Landau-Lifshitz-Navier-Stokes equation*: Can we prove the existence of the solution of the Landau-Lifshitz-Navier-Stokes equation (6.28) with physically proper boundary/initial conditions, whose solution statistics are stable (or unstable), i.e., insensitive (or sensitive) to small disturbances?

This open question essentially concerns the existence and *statistics stability* of the solution of the Landau-Lifshitz-Navier-Stokes (LLNS) equation (6.28) under small disturbances. I personally believe that it should be more difficult to give a mathematical proof for this than for the original fourth Clay millennium problem.

It should be emphasized that, in addition to thermal fluctuations, there exist many random environmental and artificial disturbances that affect turbulent flows, and which can hardly be avoided completely. In practice, such random disturbances should be considered in any mathematical model of a turbulent flow. The author therefore recommends the following *generalized Landau-Lifshitz-Navier-Stokes equation*:

$$\frac{\partial \mathbf{U}}{\partial t} + \nabla \cdot \mathbf{F}_a = \nabla \cdot \mathbf{F}_b + \nabla \cdot \mathbf{F}_s + \mathbf{F}_d, \tag{6.35}$$

where $\mathbf{F}_d$ is a random vector related to small environmental/artificial disturbances, which causes deviations to occur in the vector $\mathbf{U}(\mathbf{r}, t) = (\rho, \rho\mathbf{u}, E)^{\mathrm{T}}$ of the variables under consideration. Obviously, when $\mathbf{F}_d = 0$, then (6.35) reduces to the Landau-Lifshitz-Navier-Stokes equation (6.28). Moreover, when $\mathbf{F}_s = 0$ and $\mathbf{F}_d = 0$, the equation reduces further to the exact Navier–Stokes equation. Thus, (6.35) is more general than the LLNS equation (6.28) and the Navier–Stokes equation.

Likewise, it is also important to know the existence and statistics stability of solutions of the generalized Landau-Lifshitz-Navier-Stokes equation (6.35). Thus, the author suggests the following open question:

> **Open Question 6.3** *Existence and statistics stability of the generalized Landau-Lifshitz-Navier-Stokes equation*: Can we prove the existence of the solution of the generalized Landau-Lifshitz-Navier-Stokes equation (6.35) with physically proper boundary/initial conditions, whose statistics are stable (or unstable), i.e., insensitive (or sensitive) to small disturbances?

In practice, all three aforementioned open questions are important. Obviously, each question can be regarded as a kind of *modified fourth Clay millennium problem*. From a mathematical perspective, these open questions are challenging and comparative to the original fourth Clay millennium problem[¶]. From a physical perspective, these open questions have very important meanings and can have far-reaching repercussions.

[¶]See the website http://www.claymath.org/sites/default/files/navierstokes.pdf.

7

Periodic Orbits of the Three-Body Problem

The three-body problem [1, 19–23], i.e., the movement of three masses under Newton's laws of motion and gravitational attraction, can be traced back to Newton [18] in 1687. In 1890 Poincaré [1] was the first to discover the sensitivity dependence on initial condition (SDIC) of the trajectory of the three-body problem. SDIC became the foundation of chaos dynamics. Today, chaos dynamics is regarded as one of three greatest revolutions in physics that occurred in the 20 century, alongside quantum mechanics and Einstein's theory of relativity. Thus, the three-body problem has played a crucial role in the progress of chaos dynamics.

As pointed out by Poincaré [1], periodic orbits of the three-body system are very important, because they are "the only opening through which we can try to penetrate in a place which, up to now, was supposed to be inaccessible". Using conventional algorithms with single (or double) precision floating-point arithmetic, only three families of periodic orbits of the three-body problem with equal masses were obtained in three hundred years after Newton [18] posed the problem. This somewhat disappointing lack of progress came about primarily because the three-body system is essentially chaotic, so that, due to the butterfly-effect, it was not possible in general cases to obtain a convergent trajectory over a sufficiently long time interval by means of traditional numerical methods. Fortunately, Clean Numerical Simulation (CNS) [38–40, 43, 54] can overcome this limitation. By using CNS to gain a convergent trajectory over a sufficiently long time interval, the number of families of periodic orbits of three-body system has increased by several orders of magnitude [45–48, 52]. Moreover, Liao *et al.* [53] proposed a general roadmap to obtain periodic orbits of three-body systems with arbitrary masses by combining CNS with artificial intelligence (AI).

This chapter briefly describes how Newton's open problem was attacked by using CNS to guarantee the convergence of the trajectory over a prescribed sufficiently long time interval. It illustrates that CNS as a new powerful tool can indeed bring us something new and different.

DOI: 10.1201/9781003299622-7

7.1 Historical Review

Consider the trajectories of three point masses m_1, m_2 and m_3 whose attraction to each other is governed by Newton's gravitational law. Using Newton's second law of motion and the gravitational law, the governing equations of the N-body problem may be written,

$$m_k \frac{d^2 \mathbf{r}_k}{dt^2} = \sum_{j=1, j \neq k}^{N} \frac{G m_k m_j (\mathbf{r}_j - \mathbf{r}_k)}{|\mathbf{r}_j - \mathbf{r}_k|^3}, \quad 1 \leq k \leq N, \tag{7.1}$$

subject to the initial condition

$$\mathbf{r}_k(0) = \mathbf{r}_k^*, \quad \dot{\mathbf{r}}_k(0) = \mathbf{v}_k^*, \quad 1 \leq k \leq N, \tag{7.2}$$

where t denotes time, the dot denotes the derivative with respect to t, m_k and $\mathbf{r}_k(t)$ denote the mass and position vector of the kth body, $\mathbf{r}_k^*$ and $\mathbf{v}_k^*$ are the initial position and velocity of the kth body. If each body returns exactly to its initial state at $t = T$, such that

$$\mathbf{r}_k(T) = \mathbf{r}_k(0), \quad \dot{\mathbf{r}}_k(T) = \dot{\mathbf{r}}_k(0), \quad 1 \leq k \leq N, \tag{7.3}$$

then there exists a periodic solution of the N-body problem with period T. Newton's original three-body problem [1, 19–23] corresponds to $N = 3$ only.

7.1.1 Era of Newton, Euler, Lagrange, and Poincaré

Although Newton [18] readily derived a closed-form periodic solution of the two-body problem where $N = 2$, it proved extremely difficult to find periodic orbits for the three-body problem ($N = 3$). In fact, no periodic orbits were found until Euler reported one in 1740 and Lagrange published another in 1772. In fact, both these periodic orbits belong to the *same* family, called the "Euler-Lagrange family", according to Montgomery's topological method [19] for classifying the periodic orbits of three-body systems. Thereafter, no new periodic orbits were reported in about two centuries.

Why is the three-body problem so difficult? In 1890, Poincaré [1] pointed out that the three-body problem is *not* integrable and thus orbits in a closed-form do *not* exist for general cases. This implies that it is normally *necessary* to use numerical algorithms to solve the three-body problem. Moreover, Poincaré [1] was the first to discover the sensitivity dependence of trajectories of the three-body system on initial conditions, which became the foundation of a new field of modern science called chaotic dynamics. As mentioned earlier, Poincaré [1] also emphasized the usefulness of periodic orbits of the three-body system in enabling us to understand previously unknown facets of nonlinear systems.

7.1.2 Era of the Electronic Computer

The excellent work of Poincaré [1] represents a historical turning point in our understanding of periodic orbits of the three-body system because the non-existence of a uniform first integral of the triple system reveals the impossibility of finding its closed-form analytic solutions in general cases. This was indeed a revolutionary contribution of Poincaré [1] at a time before the invention of the electronic computer. About a half century later, Turing [115,116] published two epoch-making papers that helped form the foundation of modern computer science and artificial intelligence, and von Neumann [117] proposed the basic ideas of the so-called von Neumann machine which underpins almost all modern computer architecture.

According to Montgomery's topological method [19] for classifying periodic orbits of three-body systems, only three families of periodic orbits of the three-body system were found in 300 years after Newton, i.e.,

(1) the so-called "Euler-Lagrange family", found by Euler in 1740 and Lagrange in 1772;

(2) the so-called "BHH family", *numerically* discovered by Broucke [118, 119] in 1975, Hadjidemetriou [120] in 1975 and Hénon [121] in 1976;

(3) the so-called "Figure-eight family" of orbital trajectories of three equal masses, *numerically* discovered by Moore [122] in 1993,

until 2013 when Šuvakov and Dmitrašinović [64] *numerically* obtained eleven families of new periodic orbits of the triple system with three equal masses. Their findings strongly suggest that numerical methods offer a correct approach to find further periodic orbits of the three-body problem.

By 2013, humankind had developed supercomputers with a combined peak performance of about 1,000 petaflops, i.e., several billion billion fundamental calculations per second. Then, what prevents us from effectively finding thousands of new families of periodic orbits for the three-body problem given that such huge supercomputing power is available?

In fact, a major obstacle had already been revealed by Lorenz, who not only rediscovered the sensitivity dependence of chaotic trajectory on initial condition [2] but also popularized the concept of the "butterfly-effect" whereby a hurricane in North America might be created by the flapping of wings of a distant butterfly in South America several weeks earlier. In addition Lorenz [9,10] was the first to discover the sensitivity dependence of a chaotic trajectory on numerical algorithms using single or double precision floating-point arithmetic. Lorenz found that the use of different numerical algorithms could lead to distinctly different computer-generated trajectories of a chaotic system after a long time duration. Note that sensitivity dependence on initial conditions (SDIC) has physical meaning because any tiny difference in initial conditions has a physical meaning. However, sensitivity dependence on numerical algorithm (SDNA) has *no* physical meaning whatsoever because numerical algorithms are inherently *artificial* and the true physical solution must be *independent* of any artefact! The SDNA of a chaotic system was also confirmed by other researchers [11, 14, 32], and this unsurprisingly led to intense debate [12] with certain researchers even stating that "all chaotic responses should be simply numerical noises" and might "have nothing to do with differential equations".

The key for determining periodic orbits is to obtain convergent computer-generated trajectories of a three-body system under arbitrary initial conditions over a *sufficiently long* time interval. However, according to Poincaré [1], the three-body system is usually chaotic, and according to Lorenz [9,10], its chaotic trajectory is also sensitive to conventional numerical algorithms using single/double precision floating-point arithmetic. In practice, it is usually extremely difficult to obtain a convergent trajectory of the three-body problem over a *sufficiently* long time interval by means of conventional algorithms. This great obstacle potentially prevents prople from finding thousands of new families of periodic orbits of the three-body problem. This is exactly why one could not efficiently obtain new families of periodic orbits of the three-body system even with supercomputers whose peak performance exceeds about 1,000 petaflops!

7.2 Discovery of New Periodic Orbits by Means of CNS

To overcome the above obstacle, the author in 2009 proposed "clean numerical simulation" (CNS) [38], which can give the *convergent* trajectory of a chaotic system over a prescribed *sufficiently* long time interval. CNS offered the key advance necessary to find periodic orbits of the three-body problem, as discussed below.

Traditionally, the Runge-Kutta method with single (or double) precision floating-point arithmetic has been used to calculate the trajectory of a three-body system, with a grid search method applied to find possible candidates for the initial conditions of periodic orbits, and the Newton-Raphson iteration method [123, 124] used to progressively modify the initial condition of a possible periodic orbit. Unfortunately, the three-body problem is essentially chaotic [1] but the Runge-Kutta method with single/double precision floating-point arithmetic cannot provide the convergent trajectory of chaos over a sufficiently long time interval. Fortunately, this limitation is overcome by CNS.

In 2017, using CNS to integrate the corresponding governing equations to gain convergent trajectories of the three-body system (please see Section 4.1 for details), Li & Liao [45] successfully found 695 families of periodic planar collisionless orbits of three-body systems with three equal masses and zero

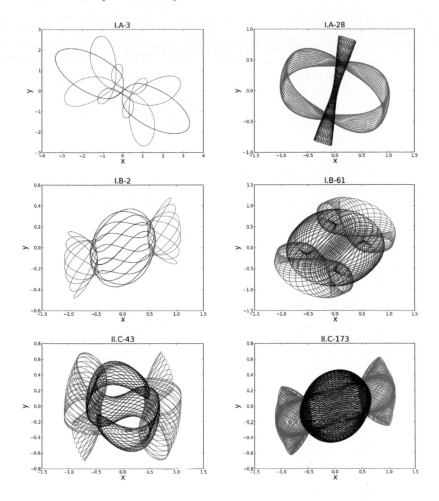

FIGURE 7.1
Periodic orbits of the three-body system with isosceles collinear configuration for a case of equal mass and zero angular momentum, discovered by Li & Liao [45] using the CNS-based strategy. Blue line: orbit of Body-1; red line, orbit of Body-2; and black line, orbit of Body-3.

angular momentum, which included Moore's Figure-eight family, the eleven families reported by Šuvakov & Dmitrašinović in 2013, and a further more than 600 new families that had never previously been reported. Here, CNS is combined with the traditional strategy: the grid search method is applied to find possible candidates for the initial conditions of periodic orbits, and Newton-Raphson iteration method [123, 124] is used to modify progressively the initial condition of a possible periodic orbit. Figure 7.1 and Figure 1.11

depict several of these new families of periodic orbits. Note that a kind of generalized Kepler's third law was revealed for the first time [45]. This indeed provides evidence of great progress, demonstrating the validity of CNS applied to discovering periodic orbits of the three-body problem.

Note that all the aforementioned families of periodic orbits of the three-body system involve bodies of the *same* masses, such that $m_1 = m_2 = m_3$. This is a rather special case from a physical viewpoint and is of purely mathematical significance. Using the same strategy, Li, Jing & Liao [46] discovered a further 1349 new families of periodic orbits of a three-body system with two equal masses and a third arbitrary mass. Similarly, Li & Liao [47] discovered a further 316 collisionless free-fall periodic orbits of a three-body system with *randomly* chosen masses, among which 313 orbits had not been previously reported. This strongly suggests that an infinite number of families of periodic orbits of the free-fall three-body system should exist, given that the mass ratio can be randomly chosen [47]. Moreover, it was found that the generalized form of Kepler's third law widely holds for the various forms of orbits that were discovered [46–48].

Note that all the new periodic orbits were obtained using the grid-search method to find a possible candidate for the initial condition of the three-body system. However, for a known periodic orbit of the three-body system with masses m_1^*, m_2^*, and m_3^*, one can find new periodic orbits of a three-body system with different masses near m_1^*, m_2^*, and m_3^* by means of the continuation method [125]. Let $\boldsymbol{u}^*$ denote a known periodic orbit when $m_1 = m_1^*, m_2 = m_2^*$, and $m_3 = m_3^*$. Taking the initial condition of $\boldsymbol{u}^*$ as a possible candidate, a new periodic orbit $\boldsymbol{u}'$ can be obtained for

$$m_1 = m_1^* + \Delta m_1, \quad m_2 = m_2^*, \quad m_3 = m_3^*$$

using the Newton-Raphson iteration method [123, 124] to progressively improve the initial condition and CNS [38–40,43,54] to determine the convergent trajectory, provided the mass increment Δm_1 is sufficiently small to guarantee iterative convergence. Similarly, a new periodic orbit can be obtained for

$$m_1 = m_1^*, \quad m_2 = m_2^* + \Delta m_2, \quad m_3 = m_3^*,$$

so long as the mass increment Δm_2 is small enough for the iterations to converge. And so on. In 2019, starting from a given periodic orbit, Li *et al.* [52] successfully discovered 135445 new periodic orbits for cases involving arbitrarily *unequal* masses, using the numerical continuation method [125] to obtain each possible candidate for the initial condition, CNS [38–40,43,54] to guarantee trajectory convergence, and Newton-Raphson iteration method [123, 124] to improve progressively the initial condition of each possible periodic orbit.

As a result, by means of the foregoing CNS-based strategy, the number of families of periodic orbits of three-body systems was increased by nearly four orders of magnitude in only four years! This provides a convincing argument for using the above-mentioned CNS-based strategy to discover many new families of periodic orbits of three-body systems. It should be emphasized that the great progress to date is primarily due to the use of CNS to guarantee trajectory convergence in three-body systems with arbitrary initial conditions, which are mostly chaotic according to Poincaré [1]. Thus, CNS offers us a key approach by which to attack Newton's famous open problem which has a history of more than three hundred years. Please note that most of these new periodic orbits had *not* been previously reported. This illustrates once again the great potential of CNS to bring something completely different and new to the study of nonlinear systems!

For more details of all the foregoing periodic orbits of the three-body system, including initial conditions, movies of the orbits, etc., please visit the websites:

A. https://github.com/sjtu-liao/three-body;

B. https://numericaltank.sjtu.edu.cn/three-body/three-body.htm.

Given the long history of the three-body problem and its fame in the scientific community, the author's discoveries of new families of periodic orbits of the three-body system have been reported twice by *New Scientist*, on 20 September 2017* and 25 May 2018†, respectively.

7.3 A Roadmap Based on CNS and Artificial Intelligence

The grid-search method is rather time-consuming. In order to increase computational efficiency, Liao *et al.* [53] utilized an Artificial Neural Network (ANN) [126–129] to determine possible candidates for the initial condition of periodic orbits of the three-body problem, using knowledge of a few known periodic orbits of the same family with various masses. ANN was chosen because it can model rather complex nonlinear relationships [128] and easily cope with classification problems when complicated boundaries are encountered [130].

Without loss of generality, let us select a known BHH orbit as the starting point to illustrate this strategy [53]. There exist many satellite periodic orbits of the BHH family of three-body systems with three equal masses [131, 132], having zero angular momentum as the initial condition, such that

$$r_1(0) = (x_1, 0), \quad r_2(0) = (1, 0), \quad r_3(0) = (0, 0), \tag{7.4}$$

---

*L. Crane, "Infamous three-body problem has over a thousand new solutions", *New Scientist* and 20 September 2017. Please visit: https://www.newscientist.com/article/2170161-watch-the-weird-new-solutions-to-the-baffling-three-body-problem/.

†C. Whyte, "Watch the weird new solutions to the baffling three-body problem", *New Scientist*, 25 May 2018; and: https://www.newscientist.com/article/2170161-watch-the-weird-new-solutions-to-the-baffling-three-body-problem/.

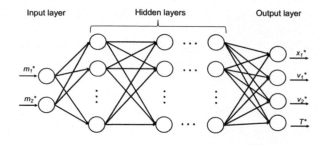

FIGURE 7.2
Artificial neural network (ANN) model. Here the input (m_1^*, m_2^*) and output $(x_1^*, v_1^*, v_2^*, T^*)$ are respectively the normalized data of masses (m_1, m_2) and the four unknown physical parameters (x_1, v_1, v_2, T).

and

$$\dot{r}_1(0) = (0, v_1), \quad \dot{r}_2(0) = (0, v_2), \quad \dot{r}_3(0) = \left(0, -\frac{(m_1 v_1 + m_2 v_2)}{m_3}\right), \qquad (7.5)$$

where r_i, $\dot{r}_i$ and m_i denote the position, velocity and mass of the i-th body, and $i = 1, 2, 3$. For simplicity, let $m_3 = 1$ in all cases. Thus, given m_1 and m_2, there are four unknown physical variables x_1, v_1, v_2, and T, where T is the time period. Note that these orbits are periodic in a frame of coordinates which rotates through an angle θ_T.

For example, let us consider a known BHH periodic orbit with initial condition

$$\begin{cases} x_1 = -1.325626981682458, \\ v_1 = -0.8933877752879044, \\ v_2 = -0.2885702941263346, \\ T = +9.199307755830397, \end{cases} \qquad (7.6)$$

and rotation angle $\theta_T = 0.383160887655628$ of the coordinate frame for $m_1 = m_2 = 1$. Taking this as a starting point and following the above CNS-based strategy [52], 36 periodic orbits with different masses in a small domain $m_1 \in [0.95, 1.00], m_2 \in [1.00, 1.05]$ (marked in green in Figure 7.3 and expressed by S_1) are obtained using the continuation method [125] (for mass increment $\Delta m_1 = \Delta m_2 = 0.01$) and Newton-Raphson iteration method [123, 124] (for modification of initial conditions of a possible periodic orbit). This provides an initial database for the ANN model.

Next, the ANN model is used to determine a relation between the input vector (m_1, m_2) and the output vector (x_1, v_1, v_2, T) of the periodic orbits. This ANN model consists of one input layer, six hidden layers, and one output layer, with 2, 1024 and 4 neurons, respectively, as shown in Figure 7.2. The optimization algorithm AMSGrad [133] is then used to minimize the mean

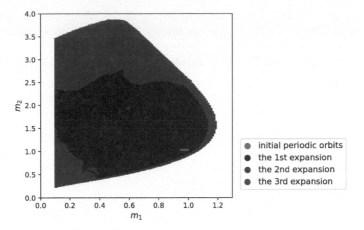

FIGURE 7.3

Mass domain (m_1, m_3) containing periodic orbits, which is enlarged progressively by extrapolation using trained ANN models. Green dots: domain S_1 occupied by initial periodic orbits; purple dots: domain S_2 determined by the first extrapolation; blue dots: domain S_3 determined by the second extrapolation; red dots: domain S_4 determined by the third extrapolation.

square error and hence train the ANN model. Initially, the results of 36 known periodic orbits are used as the training set to train the ANN model. This provides us with a relationship $\mathcal{F}_1$ between (m_1, m_2) and (x_1, v_1, v_2, T), which may be expressed by the following mapping

$$\mathcal{F}_1 : (m_1, m_2) \in S_1 \rightarrow (x_1, v_1, v_2, T), \tag{7.7}$$

that can be further applied to predict the initial conditions x_1, v_1, v_2 and period T of possible candidates for periodic orbits associated with various masses (m_1, m_2) *outside* the original domain (marked in green)

$$S_1 = \left\{ (m_1, m_2) : m_1 \in [0.95, 1.00], m_2 \in [1.00, 1.05] \right\}.$$

For any $(m_1, m_2) \in S_1$, the ANN model $\mathcal{F}_1$ defined by (7.7) can provide sufficiently accurate values of (x_1, v_1, v_2, T), corresponding to a periodic orbit. This is easy to understand. However, instead of the grid-search method, we use the ANN model $\mathcal{F}_1$ to predict possible candidates (x_1, v_1, v_2, T) for periodic orbits *outside* S_1, and apply CNS to determine the convergent trajectory over a sufficiently long time interval and the Newton-Raphson iteration method [123,124] to improve the candidate initial conditions (x_1, v_1, v_2, T) successively.

Using $\Delta m_1 = \Delta m_2 = 0.01$ and excluding those candidates in S_1, 17421 new periodic orbits were found commencing from 26364 possible candidates predicted by $\mathcal{F}_1$ within the mass regions $m_1 \in [0.1, 1.2]$ and $m_2 \in [0.4, 2.8]$ *outside* S_1, which are marked in purple in Figure 7.3 and expressed by S_2. Then, we use all the known results, i.e., 36 (in S_1) + 17421 (in S_2) = 17457 (in $S_1 \cup S_2 = \Omega_2$), as a larger training database in a similar way to obtain an enhanced relationship $\mathcal{F}_2$ between (m_1, m_2) and (x_1, v_1, v_2, T), expressed by

$$\mathcal{F}_2 : (m_1, m_2) \in \Omega_2 = S_1 \cup S_2 \rightarrow (x_1, v_1, v_2, T). \tag{7.8}$$

Similarly, using the results for (x_1, v_1, v_2, T) predicted by the ANN model $\mathcal{F}_2$ as possible candidates, 11473 new periodic orbits are found within a domain S_3 outside $\Omega_2 = S_1 \cup S_2$, which is marked in blue in Figure 7.3.

Then, all the known results, i.e., 36 (in S_1) + 17421 (in S_2) + 11473 (in S_3) = 28930 (in $S_1 \cup S_2 \cup S_3 = \Omega_3$), are then further used as an even larger training database to obtain an even better relationship $\mathcal{F}_3$ between (m_1, m_2) and (x_1, v_1, v_2, T). In other words,

$$\mathcal{F}_3 : (m_1, m_2) \in \Omega_3 = S_1 \cup S_2 \cup S_3 \rightarrow (x_1, v_1, v_2, T). \tag{7.9}$$

However, using the results for (x_1, v_1, v_2, T) predicted by $\mathcal{F}_3$ as possible candidates in a similar way, only 220 new periodic orbits are found in a domain *outside* $(m_1, m_2) \in \Omega_3 = S_1 \cup S_2 \cup S_3$, marked in red in Figure 7.3 and expressed as S_4. This suggests that the majority of periodic orbits should exist in the domain

$$S^* = S_1 \cup S_2 \cup S_3 \cup S_4. \tag{7.10}$$

Finally, all the known results, i.e., 36 (in S_1) + 17421 (in S_2) + 11473 (in S_3) + 220 (in S_4) = 29150 (in S^*), are then used as an even larger training database by which to obtain a more refined relationship $\mathcal{F}^*$ between (m_1, m_2) and (x_1, v_1, v_2, T), i.e.,

$$\mathcal{F}^* : (m_1, m_2) \in S^* \rightarrow (x_1, v_1, v_2, T), \tag{7.11}$$

where S^* is defined by Eq. (7.10). It is found that the ANN model $\mathcal{F}^*$ defined by (7.11) can provide accurate predictions of the periodic orbits for *arbitrary* values of $(m_1, m_2) \in S^*$ within a level of accuracy of 10^{-4} of the return distance (deviation). All the foregoing ANN predictions of periodic orbits are sufficiently accurate for practical use, even though the accuracy can be further modified to a specified *arbitrary* level by using the Newton-Raphson iteration method to improve the corresponding initial condition and CNS to guarantee the convergent trajectory throughout a prescribed sufficiently long period T. For example, Figure 7.4 shows six periodic orbits predicted by the ANN model $\mathcal{F}^*$ for randomly chosen masses (m_1, m_2), whose initial conditions and period T are listed in Table 7.1. Moreover, it was found [53] that all these periodic orbits in a reference frame with the same rotation angle $\theta_T = 0.38316088765562$ are linearly stable. For further details, please refer to Liao, Li and Yang [53].

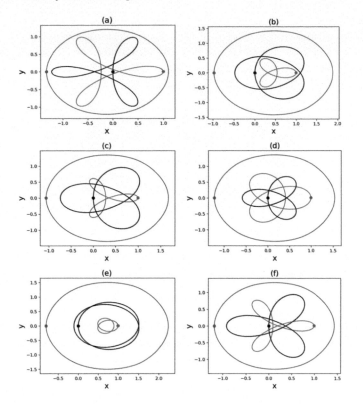

FIGURE 7.4

Relatively periodic BHH satellite periodic orbits of the three-body system with different masses (m_1, m_2) in a rotating frame of reference for rotation angle $\theta_T = 0.38316088765562$, predicted by the ANN model. Corresponding initial conditions and the period T are listed in Table 7.1. Blue line: Body-1; red line: Body-2; and black line: Body-3.

As shown in Figure 7.3, starting from the 36 periodic orbits in the rather small mass domain $(m_1, m_2) \in S_1$ (marked in green), a total of 29150 periodic orbits in the mass domain $(m_1, m_2) \in S^*$ are found by the three implementations of extrapolations/expansions using the ANN models: $\mathcal{F}_1$, $\mathcal{F}_2$, and $\mathcal{F}_3$. During this process, the mass domain (m_1, m_2) including periodic orbits enlarges from S_1 to $\Omega_2 = S_1 \cup S_2$, then to $\Omega_3 = S_1 \cup S_2 \cup S_3$, and finally to $S^* = S_1 \cup S_2 \cup S_3 \cup S_4$. Meanwhile, the corresponding ANN model, expressing the relationship between (m_1, m_2) and (x_1, v_1, v_2, T), improves from $\mathcal{F}_1$ to $\mathcal{F}_2$ to $\mathcal{F}_3$ to $\mathcal{F}^*$. In other words, the ANN model becomes wiser and wiser!

The preceding example illustrates a general road map by which to discover new periodic orbits of the three-body system (when $m_3 = 1$) of a known family,

TABLE 7.1

Initial Conditions and Period T of Relatively Periodic Orbits Predicted by the ANN Model for Six BHH Satellites with Randomly Chosen Masses $(m_1; m_2)$ (When $m_3 = 1$) under the Same Rotation Angle $\theta_T = 0.38316088765562$ of the Reference Frame with Initial Conditions Given by (7.4) and (7.5)

Case	m_1	m_2	x_1	v_1	v_2	T
(a)	1.0124	0.9968	−1.32962	−0.88963	−0.28501	9.2111
(b)	0.5312	2.2837	−0.97138	−1.37584	−0.34528	7.0421
(c)	0.8056	2.0394	−1.05795	−1.24044	−0.26723	7.4389
(d)	0.3916	0.8341	−1.21503	−1.00328	−0.53749	9.2807
(e)	0.1472	3.4219	−0.80027	−1.74811	−0.42176	5.8793
(f)	0.8413	1.4155	−1.17777	−1.05903	−0.29934	8.3444

i.e., with the same "free group element" defined by Montgomery's topological method [19]:

(1) For a three-body system with three or two equal masses, first find a set of possible candidates for the initial conditions of periodic orbits using the grid search method, and then progressively improve them using the Newton-Raphson iteration method until a sufficiently accurate periodic orbit is obtained, while using CNS to guarantee trajectory convergence.

(2) Take each known periodic orbit as a starting point from which to obtain a few new periodic orbits with different masses within a small domain (m_1, m_2) using a combination of the numerical continuation method [125] and the Newton-Raphson iteration method [123, 124]. Then, the initial conditions and periods of these new periodic orbits form an initial training set for the ANN model.

(3) For a given training set in a mass domain $(m_1, m_2) \in \Omega_k$, where $k \geq 1$ is an integer, first train the ANN model $\mathcal{F}_k$ to give accurate predictions of initial conditions and periods of the periodic orbits for various mass *within* Ω_k, say, $(m_1, m_2) \in \Omega_k$. Then, use this trained ANN model $\mathcal{F}_k$ to predict possible candidates for the initial condition and period T *out* of Ω_k so that new periodic orbits *outside* the previous mass domain $(m_1, m_2) \in \Omega_k$ can be found by again using the Newton-Raphson iteration method [123, 124] and CNS. The resulting new periodic orbits provide us with a larger training set, which could further provide us with a better trained ANN model $\mathcal{F}_{k+1}$ that can predict new periodic orbits in an even larger mass domain Ω_{k+1}. The same process can be repeated again and again so that increasing numbers of periodic orbits are found over an extending mass domain, and the trained ANN model becomes wiser and wiser, until no new periodic orbits can be found in a

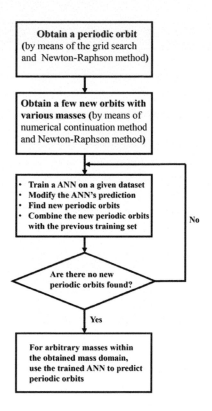

FIGURE 7.5
Roadmap for finding periodic orbits of the three-body system with arbitrary masses.

larger mass domain. Once this stage has been reached, we have created a trained ANN model $\mathcal{F}^*$ based on all known periodic orbits in the final mass domain $(m_1, m_2) \in S^*$.

(4) Randomly select hundreds or thousands of *arbitrary* masses $(m_1, m_2) \in S^*$. In each case, check whether or not the trained ANN model $\mathcal{F}^*$ gives a sufficiently accurate prediction, and also whether or not the corresponding return distance (deviation) could indeed be reduced to a tiny level such as 10^{-60}. If yes, then the trained ANN model $\mathcal{F}^*$ can provide a satisfactory prediction of periodic orbits in the mass domain $(m_1, m_2) \in S^*$.

The above roadmap has general applicability; its pipeline is summarized in Figure 7.5. For more details, please refer to Liao, Li and Yang [53].

7.4   Scientific Significance of Discovering New Periodic Orbits

Even though the famous three-body problem [1,19–23] was first suggested by Newton [18] in 1687, the discovery of new periodic orbits of the three-body system with arbitrary masses still remains open, even in the 2010s when supercomputers became available with peak performance about 1,000 petaflops, i.e., several billion billion fundamental calculations per second. The problem's open nature persists primarily because a three-body system is essentially chaotic, with trajectories that are sensitive not only to the initial condition [1], i.e., the butterfly-effect [2], but also to artificial factors such as the numerical algorithm and time-step [9,10]. Therefore, convergent trajectories of the three-body system over a sufficiently long time interval are almost impossible to obtain using traditional algorithms in single/double precision floating-point arithmetic. This is the key reason why only three families of periodic orbits of three-body problem were found within more than 300 years after Newton [18] first posed the problem in 1687.

This book has demonstrated how the above great obstacle can be completely overcome by means of CNS, which can provide the convergent trajectory of a triple system containing arbitrary masses over a prescribed sufficiently long time interval! Combining CNS with the traditional grid-search method to find possible candidates for initial conditions and the Newton-Raphson iteration method [123,124] to progressively improve the initial conditions, one can gain thousands upon thousands of new families of the three-body system by means of CNS [45–47]. Besides, given any known family of periodic orbits, one can combine CNS with the continuation numerical method (instead of the grid search method) and Newton-Raphson iteration method [123,124] to discover new periodic orbits for continuous variations in the masses [52,53]. Furthermore, given a few known periodic orbits of a family of the three-body system, one can use techniques related to artificial intelligence such as the Artificial Neural Network (ANN) [126–129] to efficiently predict candidates of initial conditions for possible periodic orbits of the same family with various masses [53]. This provides us with a general roadmap by which to discover new periodic orbits of three-body systems. It should be emphasized that, although all known periodic orbits of triple-body systems are planar, i.e., in two-dimensions, theoretically speaking, the same strategy described in this chapter can be applied to discover periodic orbits of N-body system in three-dimensions, where $N \geq 3$. It would be wonderful if three-dimensional stable periodic orbits of N-bodies (with a large N, such as $N = 7$ or 10) could be discovered in the near future.

Today, combining CNS with other numerical techniques, especially those related to artificial intelligence such as Artificial Neural Networks (ANNs), *nothing* can prevent us from discovering an immense number of new families of periodic orbits of the three-body system with arbitrary masses! This could lead to an enormous database being established of the periodic orbits of triple systems, which could enrich and deepen our understanding of the famous three-body problem from a theoretical standpoint, noting (as mentioned before) that such trajectories comprise "the only opening through which we can try to penetrate in a place which, up to now, was supposed to be inaccessible" [1]. Hopefully, certain physical laws, such as a generalized Kepler's law [45] governing periodic orbits of triple-body systems with *arbitrary* masses, could be discovered by using artificial intelligence to analyze the huge number of orbits contained within the database.

In practice, all observed periodic triple-body systems are hierarchical, i.e., similar to the Sun, Earth and Moon. It is widely believed that non-hierarchical triple systems are unstable. However, Li *et al.* [52] discovered 13,315 stable *non-hierarchical* periodic orbits (among a total of 135,445 non-hierarchical periodic orbits), some of which had mass ratios close to those of hierarchical triple systems observed by astronomers. Hopefully, a few complicated (hierarchical or non-hierarchical) periodic orbits of actual three-body systems with distinctly unequal masses will be discovered in the future by comparing observed astronomical data with periodic orbits in the above-mentioned massive database.

It is true that a numerical periodic orbit is not a close-form solution. However, from viewpoints of mathematics, using the above-mentioned CNS-based strategy with a sufficiently large number N_s of significant digits of multiple precision floating-point arithmetic, one can gain periodic orbits of the three-body problem in a required *arbitrary* accuracy. From viewpoints of physics, the so-called Planck length

$$l_P = \sqrt{\frac{\hbar\,G}{c^3}} \approx 1.616252(81) \times 10^{-35} \quad (\text{m}) \tag{7.12}$$

is the length scale at which all quantum mechanics, gravity and relativity interact rather strongly, where $\hbar$ is the reduced Planck's constant, G is the gravitational constant, and c is the speed of light in a vacuum. According to string theory [76], the Planck length l_P is the order of magnitude of oscillating strings that form elementary particles, and *a shorter length does not make physical sense*. Notably, in some forms of quantum gravity, it becomes *impossible* to determine the difference between two locations *less* than one Planck length apart. So, from physical viewpoints, it is *unnecessary* for the spatial position of a periodic orbit to be accurate at a level lower than one Planck length $l_P = 1.616252 \times 10^{-35}$ meter, corresponding to a non-dimensional length 1.8×10^{-56} if the diameter of the Milky Way Galaxy is used as the characteristic length. Using the CNS-base strategy, Liao *et al.* [53] successfully obtained periodic orbits of the three-body problem with the initial condition and period in accuracy of one hundred significant digits, which certainly can be regarded as "exact" solution of the three-body problem from physical viewpoints.

At the time of writing, the N-body problem [134] where N is very large (such as $N = 512^3$ or $N = 1024^3$) has been modelled by means of numerical algorithms based on single (or double) precision arithmetic. However, according to the groundbreaking work of Poincaré [1], it is quite likely that such a N-body system is chaotic, and so numerical noise δ' affecting the non-periodic trajectory given by conventional algorithms exponentially increases so that the numerical simulation $\mathcal{S}'$ is a mixture, i.e., $\mathcal{S}' = \mathcal{P} + \delta'$, where $\mathcal{P}$ is the true physical solution and δ' is numerical noise, which are mostly of the same order of magnitude, say, $\delta' \sim \mathcal{P} \sim \mathcal{S}'$. Therefore, it would be highly desirable to investigate in future whether or not the statistical results of such N-body systems are stable, i.e., insensitive to small disturbances, mathematically speaking,

$$\langle \mathcal{P} + \delta' \rangle = \langle \mathcal{P} \rangle, \quad \text{when } \delta' \sim \mathcal{P} \text{ mostly,} \tag{7.13}$$

where $\langle \rangle$ denotes a statistical operator. To the best of the author's knowledge, there are no rigorous proofs available to date about the statistics stability (7.13) of the general N-body problem. So, I recommend the following open question:

Open question 7.1 *Statistics stability of N-body problem*: Can we rigorously prove (or disprove) the statistics stability $\langle \mathcal{P} + \delta' \rangle = \langle \mathcal{P} \rangle$ of the N-body problem with physically proper initial conditions when $\delta' \sim \mathcal{P}$ mostly, and N is a quite large positive integer such as $N = 512^3$ or $N = 1024^3$?

Note that high performance computing and artificial intelligence have recently played very important roles in the discovery of new periodic orbits of Newton's famous three-body problem. The development of these tools is due to contributions by many mathematicians, scientists, and engineers since the late 1600s! Particular mention should be made of four pioneers. First, Poincaré [1], who founded chaos dynamics and brought about a historical turning point by proving the non-existence of the uniform first integral of the triple-body system, which implies the need to use numerical methods to solve general cases. Second, Turing [115,116], whose papers became the foundation of modern computers and artificial intelligence. Third, Von Neumann [117], who proposed the concept of so-called "Von Neumann - Machine" that underpins modern computers. And fourth, Jack S. Kilby, winner of the Nobel prize for physics in 2000, who contributed to the invention of the integrated circuit.

To conclude, the famous Newton's three-body problem has provided us with an excellent test bed by which to examine new theories accompanied by the invention of new tools that enable us to gain a much better understanding of our world and the universe in which it travels. Of the resulting tools, clean numerical simulation (CNS) offers great opportunities for future exploration of chaos, turbulence and so on.

Bibliography

[1] J.H. Poincaré. Sur le probléme des trois corps et les équations de la dynamique. Divergence des séries de m. Lindstedt. *Acta Mathematica*, 13:1–270, 1890.

[2] E.N. Lorenz. Deterministic non-periodic flow. *Journal of the Atmospheric Sciences*, 20(2):130–141, 1963.

[3] T.Y. Li and J.A. Yorke. Period three implies chaos. *The American Mathematical Monthly*, 82:985–992, 1975.

[4] M.J. Feigenbaum. Quantitative universality for a class of non-linear transformations. *Journal of Statistical Physics*, 19(1):25–52, 1978.

[5] E.N. Lorenz. *The Essence of Chaos*. University of Washington Press, Seattle, 1993.

[6] Yoshisuke Ueda. *The Road to Chaos*. Aerial Pr, 1993.

[7] P. Smith. *Explaining Chaos*. Cambridge University Press, Cambridge, 1998.

[8] J.C. Sprott. *Elegant Chaos: Algebraically Simple Chaotic Flows*. World Scientific, Singapore, 2010.

[9] E.N. Lorenz. Computational chaos - a prelude to computational instability. *Physica D*, 15:299–317, 1989.

[10] E.N. Lorenz. Computational periodicity as observed in a simple system. *Tellus-A*, 58:549–559, 2006.

[11] J. Teixeira, C.A. Reynolds, and K. Judd. Time step sensitivity of nonlinear atmospheric models: Numerical convergence, truncation error growth, and ensemble design. *Journal of the Atmospheric Sciences*, 64:175–188, 2007.

[12] L.S. Yao and D. Hughes. Comment on "computational periodicity as observed in a simple system" by Edward N. Lorenz (2006). *Tellus-A*, 60:803–805, 2008.

[13] E.N. Lorenz. Reply to comment by L.-S. Yao and D. Hughes. *Tellus-A*, 60:806–807, 2008.

[14] W.G. Hoover and C.G. Hoover. Comparison of very smooth cell-model trajectories using five symplectic and two Runge-Kutta integrators. *Computational Methods in Science and Technology*, 21:109–116, 2015.

[15] Cixin Liu. *The Three Body (1) – Remembrance of Earth's Past*. GOMO, 2008.

[16] S. Orszag. Numerical methods for the simulation of turbulence. *The Physics of Fluids*, 12:250–257, 1969.

[17] S. Orszag. Analytical theories of turbulence. *Journal of Fluid Mechanics*, 41:363–386, 1970.

[18] I. Newton. *Philosophiae naturalis principia mathematica (Mathematical Principles of Natural Philosophy)*. London, Royal Society Press, 1687.

[19] R. Montgomery. The N-body problem, the braid group, and action-minimizing periodic solutions. *Nonlinearity*, 11(2):363, 1998.

[20] R. Montgomery. The zero angular momentum, three-body problem: all but one solution has syzygies. *Ergodic Theory and Dynamical Systems*, 27(6):1933–1946, 2007.

[21] P.P. Iasko and V.V. Orlov. Search for periodic orbits in the general three-body problem. *Astronomy Reports*, 58(11):869–879, Nov 2014.

[22] A.M. Archibald, N.V. Gusinskaia, J.W.T. Hessels, A.T. Deller, D.L. Kaplan, D.R. Lorimer, R.S. Lynch, S.M. Ransom, and I.H. Stairs. Universality of free fall from the orbital motion of a pulsar in a stellar triple system. *Nature*, 559(7712):73–76, 2018.

[23] N.C. Stone and N.W.C. Leigh. A statistical solution to the chaotic, non-hierarchical three-body problem. *Nature*, 576(7787):406–410, 2019.

[24] T.S. Parker and L.O. Chua. *Practical Numerical Algorithms for Chaotic Systems*. Springer-Verlag, New York, 1989.

[25] H.G. Schuster and W. Just. *Deterministic Chaos: An Introduction*. Wiley-VCH Verlag GmbH & Co. KGaA, Weinheim, 2005. 4th edition.

[26] E. Forest and R.D. Ruth. Fourth-order symplectic integration. *Physica D: Nonlinear Phenomena*, 43(1):105–117, 1990.

[27] H. Yoshida. Construction of higher order symplectic integrators. *Physics Letters A*, 150(5):262–268, 1990.

[28] J. Laskar and P. Robutel. High order symplectic integrators for perturbed Hamiltonian systems. *Celestial Mechanics and Dynamical Astronomy*, 80(1):39–62, 2001.

[29] H. Qin and X. Guan. Variational symplectic integrator for long-time simulations of the guiding-center motion of charged particles in general magnetic fields. *Physical Review Letters*, 100:035006, 2008.

[30] A. Farrés, J. Laskar, S. Blanes, F. Casas, J. Makazaga, and A. Murua. High precision symplectic integrators for the solar system. *Celestial Mechanics and Dynamical Astronomy*, 116(2):141–174, 2013.

[31] R.I. McLachlan, K. Modin, and O. Verdier. Symplectic integrators for spin systems. *Physical Review E*, 89(6):061301, 2014.

[32] J.P. Li, Q.C. Zeng, and J.F. Chou. Computational uncertainty principle in nonlinear ordinary differential equations (I) numerical results. *Science China - Technological Sciences*, 44(1):55–74, 2001.

[33] A. Gibbon. A program for the automatic integration of differential equations using the method of Taylor series. *The Computer Journal*, 3:108, 1960.

[34] D. Barton, I.M. Willers, and R.V.M. Zahar. The automatic solution of systems of ordinary differential equations by the method of Taylor series. *The Computer Journal*, 14(3):243–248, 1971.

[35] G. Corliss and D. Lowery. Choosing a stepsize for Taylor series methods for solving ODEs. *Journal of Computational and Applied Mathematics*, 3(4):251–256, 1977.

[36] G.F. Corliss and Y.F. Chang. Solving ordinary differential equations using Taylor series. *ACM Transactions on Mathematical Software*, 8:114–144, 1982.

[37] L.S. Yao. Computed chaos or numerical errors. *Nonlinear Analysis: Modelling and Control*, 15:109 126, 2010.

[38] Shijun Liao. On the reliability of computed chaotic solutions of nonlinear differential equations. *Tellus A*, 61(4):550–564, 2009. (https://doi.org/10.1111/j.1600-0870.2009.00402.x).

[39] Shijun Liao. On the numerical simulation of propagation of micro-level inherent uncertainty for chaotic dynamic systems. *Chaos, Solitons & Fractals*, 47:1–12, 2013. (https://doi.org/10.1016/j.chaos.2012.11.009).

[40] Shijun Liao. Physical limit of prediction for chaotic motion of three-body problem. *Communications in Nonlinear Science and Numerical Simulation*, 19(3):601–616, 2014. (https://doi.org/10.1016/j.cnsns.2013.07.008).

[41] Shijun Liao and Peifeng Wang. On the mathematically reliable long-term simulation of chaotic solutions of Lorenz equation in the interval [0,10000]. *Science China - Physics, Mechanics & Astronomy*, 57(2):330–335, 2014. (https://doi.org/10.1007/s11433-013-5375-z).

[42] Shijun Liao and Xiaoming Li. On the inherent self-excited macroscopic randomness of chaotic three-body systems. *International Journal of*

Bifurcation and Chaos, 25(9):1530023, 2015. (`https://doi.org/10.1142/S0218127415300232`).

[43] Shijun Liao. On the clean numerical simulation (CNS) of chaotic dynamic systems. *Journal of Hydrodynamics*, 29(5):729–747, 2017. (`https://doi.org/10.1016/S1001-6058(16)60785-0`).

[44] Zhiliang Lin, Lipo Wang, and Shijun Liao. On the origin of intrinsic randomness of Rayleigh-Bénard turbulence. *Science China Physics, Mechanics & Astronomy*, 60(1):14712, 2017. (`https://doi.org/10.1007/s11433-016-0401-5`).

[45] Xiaoming Li and Shijun Liao. More than six hundred new families of Newtonian periodic planar collisionless three-body orbits. *Science China - Physics Mechanics & Astronomy*, 60(12):129511, 2017. (`https://doi.org/10.1007/s11433-017-9078-5`).

[46] Xiaoming Li, Yipeng Jing, and Shijun Liao. Over a thousand new periodic orbits of a planar three-body system with unequal masses. *Publications of the Astronomical Society of Japan*, 70(4):64, 2018. (`https://doi.org/10.1093/pasj/psy057`).

[47] Xiaoming Li and Shijun Liao. Collisionless periodic orbits in the free-fall three-body problem. *New Astronomy*, 70:22–26, 2019. (`https://doi.org/10.1016/j.newast.2019.01.003`).

[48] Shijun Liao and Xiaoming Li. On the periodic solutions of the three-body problem. *National Science Review*, 6:1070–1071, 2019. (`https://doi.org/10.1093/nsr/nwz102`).

[49] Tianli Hu and Shijun Liao. On the risks of using double precision in numerical simulations of spatiotemporal chaos. *Journal of Computational Physics*, 418:109629, 2020. (`https://doi.org/10.1016/j.jcp.2020.109629`).

[50] Shijie Qin and Shijun Liao. Influence of numerical noises on computer-generated simulation of spatiotemporal chaos. *Chaos, Solitons and Fractals*, 136:109790, 2020. (`https://doi.org/10.1016/j.chaos.2020.109790`).

[51] Tianzhuang Xu and Shijun Liao. Accurate predictions of chaotic motion of a free fall disk. *Phys. Fluids*, 33:037111, 2021. (`https://doi.org/10.1063/5.0039688`).

[52] Xiaoming Li, Xiaochen Li, and Shijun Liao. One family of 13315 stable periodic orbits of non-hierarchical unequal-mass triple systems. *Science China - Physics Mechanics & Astronomy*, 64(1):219511, 2021. (`https://doi.org/10.1007/s11433-020-1624-7`).

[53] Shijun Liao, Xiaoming Li, and Yu Yang. Three-body problem – from Newton to supercomputer plus machine learning. *New Astron-*

omy, 96:101850, 2022. (`https://doi.org/10.1016/j.newast.2022.101850`).

[54] Shijun Liao and Shijie Qin. Ultra-chaos: An insurmountable objective obstacle of reproducibility and replication. *Advances in Applied Mathematics and Mechanics*, 14(4):799–815, 2022. (`https://doi.org/10.4208/aamm.OA-2021-0364`).

[55] Shijie Qin and Shijun Liao. Large-scale influence of numerical noises as artificial stochastic disturbances on a sustained turbulence. *Journal of Fluid Mechanics*, 948:A7, 2022. (`https://doi.org/10.1017/jfm.2022.710`).

[56] Yu Yang, Shijie Qin, and Shijun Liao. Ultra-chaos of a mobile robot: A higher disorder than normal-chaos. *Chaos, Solitons and Fractals*, 167:113037, 2023. (`https://doi.org/10.1016/j.chaos.2022.113037`).

[57] Shijie Qin and Shijun Liao. A kind of Lagrangian chaotic property of the Arnold-Beltrami-Childress flow. *Journal of Fluid Mechanics*, 960:A15, 2023. (`https://doi.org/10.1017/jfm.2023.190`).

[58] Shijie Qin and Shijun Liao. A self-adaptive algorithm of the clean numerical simulation (CNS) for chaos. *Advances in Applied Mathematics and Mechanics*, 15 (5):1191–1215, 2023. (`https://doi.org/10.4208/aamm.OA-2022-0340`).

[59] D. Monniaux. The pitfalls of verifying floating-point computations. *ACM Transactions on Programming Languages and Systems*, 30(3):1–41, 2008.

[60] Office of U. S. Government Accountability. Patriot missile defense: Software problem led to system failure at Dhahran, Saudi Arabia. *Technical Report GAO/IMTEC-92-26 (1992)*, 1992.

[61] B. Fornberg. *A Practical Guide to Pseudospectral Methods.* Cambridge University Press, Cambridge, UK, 1996.

[62] P. Oyanarte. MP – a multiple precision package. *Computer Physics Communications*, 59(2):345–358, 1990.

[63] O.E. Rössler. An equation for hyperchaos. *Physics Letters A*, 71(2–3):155–157, 1979.

[64] Milovan Šuvakov and V. Dmitrašinović. Three classes of Newtonian three-body planar periodic orbits. *Physical Review Letters*, 110:114301, Mar 2013.

[65] L. Crane. Infamous three-body problem has over a thousand new solutions. *New Scientist*, 20 September 2017.

[66] C. Whyte. Watch the weird new solutions to the baffling three-body problem. *New Scientist*, 25 May 2018.

[67] R. Barrio, F. Blesa, and M. Lara. VSVO formulation of the Taylor method for the numerical solution of ODEs. *Computers & Mathematics with Applications*, 50(1):93–111, 2005.

[68] Á. Jorba and M. Zou. A software package for the numerical integration of ODEs by means of high-order Taylor methods. *Experimental Mathematics*, 14(1):99–117, 2005.

[69] A. Abad, R. Barrio, F. Blesa, and M. Rodriguez. Algorithm 924: TIDES, a Taylor series integrator for differential equations. *ACM Transactions on Mathematical Software*, 39(1):1–28, 2012.

[70] R. Barrio, A. Denab, and W. Tucker. A database of rigorous and high-precision periodic orbits of the Lorenz model. *Computer Physics Communications*, 194:76–83, 2015.

[71] P.F. Wang, J.P. Li, and Q. Li. Computational uncertainty and the application of a high-performance multiple precision scheme to obtaining the correct reference solution of Lorenz equations. *Numer Algorithms*, 59:147–159, 2012.

[72] J.W. Cooley and J.W. Tukey. An algorithm for the machine calculation of complex Fourier series. *Mathematics of Computation*, 19(90):297–301, 1965.

[73] J.B. Keller, D.W. Mclaughlin, and G.C. Papanicolaou. *Surveys in Applied Mathematics*. Plenum Press, 1995.

[74] R. Chacón, A. Bellorín, L.E. Guerrero, and J.A. González. Spatiotemporal chaos in sine-Gordon systems subjected to wave fields: Onset and suppression. *Physical Review E*, 77(4):046212, 2008.

[75] M.A. Ferré, M.G. Clerc, S. Coulibally, R.G. Rojas, and M. Tlidi. Localized structures and spatiotemporal chaos: comparison between the driven damped sine-Gordon and the Lugiato-Lefever model. *The European Physical Journal D*, 71(6):172, 2017.

[76] J. Polchinski. *String Theory*. Cambridge University Press, Cambridge, 1998.

[77] W. Heisenberg. Über den anschaulichen Inhalt der quantentheoretischen Kinematik und Mechanik. *Zeitschrift für Physik*, 43(3–4):172 –198, 1927.

[78] B. Saltzman. Finite amplitude free convection as an initial value problem – I. *Journal of Atmospheric Sciences*, 19:329, 1962.

[79] A.V. Getling. *Rayleigh-Bénard Convection: Structures and Dynamics*. World Scientific, 1998.

[80] M. Wu, G. Ahlers, and D.S. Cannell. Thermally induced fluctuations below the onset of Rayleigh-Bénard convection. *Physical Review Letters*, 75:1743, 1995.

[81] G. Ahlers and Jaechul Oh. Critical phenomena near bifurcations in non-equilibrium systems. *International Journal of Modern Physics B*, 17(22):3899–3907, 2003.

[82] J. Wang, Q. Li, and W. Ee. Study of the instability of the Poiseuille flow using a thermodynamic formalism. *Proceeding of the National Academy of Sciences*, 112(31):9518–9523, 2015.

[83] A.I. Khinchin. *Mathematical Foundations of Statistical Mechanics*. Dover Publications, 1949.

[84] L.D. Landau. On the problem of turbulence (in Russian). *Doklady Akad. Nauk SSSR*, 44:339, 1944. English translation in: D. Ter Hair (editor), *Collected papers of L.D. Landau* (Pergamon, Oxford, 1965) pp. 387-391.

[85] A. Tsinober. *An Informal Conceptual Introduction to Turbulence*. Springer, Dordrecht, 2009.

[86] N. Stankevich, A. Kazakov, and S. Gonchenko. Scenarios of hyperchaos occurrence in 4D Rössler system. *Chaos*, 30:123129, 2020.

[87] T. Kapitaniak, K.E. Thylwe, I. Cohen, and J. Wjewoda. Chaos - hyperchaos transition. *Chaos, Solitons & Fractals*, 5(10):2003–2011, 1995.

[88] G. Baier and S. Sahle. Design of hyperchaotic flows. *Physical Review E*, 51:2712–2714, 1995.

[89] H. Broer, C. Simo, and R. Vitolo. Bifurcations and strange attractors in the Lorenz-84 climate model with seasonal forcing. *Nonlinearity*, 15(4):1205, 2002.

[90] V.I. Arnold. Sur la topologie des écoulements stationnaires des fluides parfaits. *Comptes rendus de l'Académie des Sciences*, 261:17–20, 1965.

[91] T. Dombre, U. Frisch, J.M. Greene, M. Hénon, A. Mehr, and A.M. Soward. Chaotic streamlines in the ABC flows. *Journal of Fluid Mechanics*, 167:353–391, 1986.

[92] D. Galloway and U. Frisch. Dynamo action in a family of flows with chaotic streamlines. *Geophysical & Astrophysical Fluid Dynamics*, 36(1):53–83, 1986.

[93] D. Galloway and U. Frisch. A note on the stability of a family of space-periodic Beltrami flows. *Journal of Fluid Mechanics*, 180:557–564, 1987.

[94] A.A. Didov and M.Y. Uleysky. Analysis of stationary points and their bifurcations in the ABC-flow. *Applied Mathematics and Computation*, 330:56–64, 2018.

[95] A.A. Didov and M.Y. Uleysky. Nonlinear resonances in the ABC-flow. *Chaos: An Interdisciplinary Journal of Nonlinear Science*, 28(1):013123, 2018.

[96] O. Podvigina and A. Pouquet. On the non-linear stability of the 1: 1: 1 ABC flow. *Physica D*, 75(4):471–508, 1994.

[97] S.B. Pope. *Turbulent Flows*. IOP Publishing, 2001.

[98] L. Kuznetsov and G.M. Zaslavsky. Passive particle transport in three-vortex flow. *Physical Review E*, 61(4):3777–3792, 2000.

[99] G.D. Birkhoff. Proof of the ergodic theorem. *Proceedings of the National Academy of Sciences USA*, 17(12):656–660, 1931.

[100] J.V. von Neumann. Proof of the quasi-ergodic hypothesis. *Proceedings of the National Academy of Sciences USA*, 18(1):70–82, 1932.

[101] C.C. Moore. Ergodic theorem, ergodic theory, and statistical mechanics. *Proceedings of the National Academy of Sciences USA*, 112(7):1907–1911, 2015.

[102] M. Baker and D. Penny. Is there a reproducibility crisis? *Nature*, 533, 2016.

[103] F.C. Camerer and *et al.* Evaluating replicability of laboratory experiments in economics. *Science*, 351(6280):1433, 2016.

[104] R.D. Peng. Reproducible research in computational science. *Science*, 334:1226, 2011.

[105] O. Mesnard and L.A. Barba. Reproducible and replicable computational fluid dynamics: It is harder than you think. *Computing in Science & Engineering*, 19(4):44–55, 2017.

[106] D.J. Benjamin and *et al.* Redefine statistical significance. *Nature Human Behaviour*, 2:6–10, 2018.

[107] K. Sugiyama, E. Calzavarini, S. Grossmann, and D. Lohse. Flow organization in two-dimensional non-Oberbeck-Boussinesq Rayleigh-Bénard convection in water. *Journal of Fluid Mechanics*, 637:105–135, 2009.

[108] Y. Zhang, Q. Zhou, and C. Sun. Statistics of kinetic and thermal energy dissipation rates in two-dimensional turbulent Rayleigh-Bénard convection. *Journal of Fluid Mechanics*, 814:165–184, 2017.

[109] R.M. McMullen, M.C. Krygier, J.R. Torczynski, and M.A. Gallis. Navier-Stokes equations do not describe the smallest scales of turbulence in gases. *Physical Review Letters*, 128(11):114501, 2022.

[110] L.D. Landau and E.M. Lifshitz. *Fluid Mechanics, volume 6 of Course of Theoretical Physics*. Pergamon, 1959.

[111] R. Graham. Hydrodynamic fluctuations near the convection instability. *Physical Review A*, 10(5):1762, 1974.

[112] J. Swift and P.C. Hohenberg. Hydrodynamic fluctuations at the convective instability. *Physical Review A*, 15(1):319, 1977.

[113] J.B. Bell, A.L. Garcia, and S.A. Williams. Numerical methods for the stochastic Landau-Lifshitz Navier-Stokes equations. *Physical Review E*, 76(1):016708, 2007.

[114] A. Donev, E. Vanden-Eijnden, A. Garcia, and J. Bell. On the accuracy of finite-volume schemes for fluctuating hydrodynamics. *Communications on Applied Mathematics and Computation*, 5(2):149–197, 2010.

[115] A. Turing. On computable numbers, with an application to the Entscheidungs problem. *Proceedings of the London Mathematical Society*, 42:230–265, 1936.

[116] A. Turing. Computing machinery and intelligence. *Mind*, 50:433–460, 1950.

[117] J. Von Neumann. *The Computer and the Brain*. Yale University Press, 1958.

[118] R. Broucke. On relative periodic solutions of the planar general three-body problem. *Celestial Mechanics*, 12(4):439–462, 1975.

[119] R. Broucke and D. Boggs. Periodic orbits in the planar general three-body problem. *Celestial Mechanics*, 11(1):13–38, 1975.

[120] J.D. Hadjidemetriou. The stability of periodic orbits in the three-body problem. *Celestial Mechanics*, 12(3):255–276, 1975.

[121] M. Hénon. A family of periodic solutions of the planar three-body problem, and their stability. *Celestial Mechanics*, 13(3):267–285, 1976.

[122] C. Moore. Braids in classical dynamics. *Physical Review Letters*, 70:3675–3679, Jun 1993.

[123] C.F. Stavros. Methods for locating periodic orbits in highly unstable systems. *Journal of Molecular Structure: THEOCHEM*, 341(1):91–100, 1995.

[124] M. Lara and J. Pelaez. On the numerical continuation of periodic orbits - an intrinsic, 3-dimensional, differential, predictor-corrector algorithm. *Astronomy and Astrophysics*, 389(2):692–701, 2002.

[125] E.L. Allgower and K. Georg. *Introduction to Numerical Continuation Methods*, volume 45. SIAM, 2003.

[126] R. Andrews, J. Diederich, and A.B. Tickle. Survey and critique of techniques for extracting rules from trained artificial neural networks. *Knowledge-Based Systems*, 8(6):373–389, 1995.

[127] M. Gevrey, I. Dimopoulos, and S. Lek. Review and comparison of methods to study the contribution of variables in artificial neural network models. *Ecological Modelling*, 160(3):249–264, 2003.

[128] D.J. Livingstone. *Artificial Neural Networks: Methods and Applications*. Springer, 2008.

[129] O.I. Abiodun, A. Jantan, A.E. Omolara, K.V. Dada, N.A. Mohamed, and H. Arshad. State-of-the-art in artificial neural network applications: A survey. *Heliyon*, 4(11):00938, 2018.

[130] R. Bala and D. Kumar. Classification using ann: A review. *Int. J. Comput. Intell. Res*, 13(7):1811–1820, 2017.

[131] M.R. Janković and V. Dmitrašinović. Angular momentum and topological dependence of Kepler's third law in the Broucke-Hadjidemetriou-Hénon family of periodic three-body orbits. *Physical Review Letters*, 116:064301, 2016.

[132] M.R. Janković and V. Dmitrašinović. A guide to hunting periodic three-body orbits with non-vanishing angular momentum. *Computer Physics Communications*, 250:107052, 2020.

[133] S.J. Reddi, S. Kale, and S. Kumar. On the convergence of Adam and beyond. *arXiv preprint:1904.09237*, 2019.

[134] A. Marciniak. *The General N-body Problem*. In: Numerical Solutions of the N-Body Problem. Mathematics and Its Applications (East European Series, vol 19). Springer, Dordrecht, 1985.

Index